Risk and Reward

N. Richard Werthamer

Risk and Reward

The Science of Casino Blackjack

Second Edition

 Springer

N. Richard Werthamer
Chelsea Technologies
Sag Harbor, NY, USA

ISBN 978-3-030-08240-6   ISBN 978-3-319-91385-8   (eBook)
https://doi.org/10.1007/978-3-319-91385-8

Printed on acid-free paper

This Springer imprint is published by the registered company Springer International Publishing AG part of Springer Nature.
The registered company address is: Gewerbestrasse 11, 6330 Cham, Switzerland

Preface

How did a physicist like me get involved with blackjack? I skied in. As a young doctoral candidate at the University of California, Berkeley, I would occasionally join my friends on ski trips to the Lake Tahoe area. Always on a student's tight budget, we found that we could stay inexpensively in Reno motels and that the cheapest meals were at the coffee shops astutely placed at the back of the town's casinos. I found it hard to pass all the gambling frenzy without joining in. Fortunately for that budget and my financial peace of mind, I was usually lucky in winning a few dollars. Once, I paid for my entire trip with an unlikely hit at roulette.

When I learned that an optimized method for playing blackjack (what is now called Basic Strategy) had been published in a scholarly statistics journal, I quickly looked it up in the university library. Blackjack became my game of choice for subsequent Tahoe excursions and, in later years after skiing and I had parted company, for trips to other destinations with casinos. But my real recreation of choice became blackjack analysis!

My original motive for exploring its mathematics was to see if the articles and books I had chanced upon were correct, or if I could find a better way to win. Visions of big money danced in my head! Later, as I came to appreciate the realities involved, my interest shifted to the mathematics for its own sake, complex enough to be a challenge even for the professional scientist I had become yet easy enough to yield to a sustained effort on almost every issue. Also, I wanted to verify (or, as sometimes happened, falsify) claims about the best way to play, asserted in the literature with little or no proof.

When I had finally addressed all the questions that occurred to me, my first instinct as a scientist was to write up the results and publish them in a suitable scholarly journal: an archival record for other practitioners of applied math. Yet I came to realize that the results were sufficiently insightful about actually playing blackjack in the real world that they would be even more valuable to the wider casino-going public. To reconcile my internal debate between these two readerships, I decided to serve both at once. The outcome is this single two-part book, which first describes all its conclusions in completely nonmathematical terms for the many who enjoy gambling and then also provides detailed derivations of those conclusions so that they can be followed and checked by the more mathematically inclined. Some of the derivations rely only on algebra; others take varying levels of calculus and/or probability theory.

When talking to friends about my blackjack project, they invariably ask whether I have "tested my theories in a casino." I respond that the conclusions drawn from a mathematical proof are intrinsically true and don't need field testing to be valid. I suggest that they should rather be asking whether I have "applied" the theories, and I add that analyzing blackjack is at least as entertaining for me as playing it. I also point out that the cash outcome of playing several hundred hands at an actual blackjack table, which would occupy a full evening, is little more than an unpredictable statistical fluctuation rather than a test of the methods. Many millions of hands, simulated on a computer, are needed to approach significant conclusions. I'm happy to have examined the game in comprehensive detail, without having risked a single dime—even though I haven't won one, either!

Several leading authorities on blackjack have been generous in sharing their time and knowledge with me. Early on, Stanford Wong and Anthony Curtis were encouraging and introduced me to several other prominent figures. One of them, Don Schlesinger, intensively reviewed more than one preliminary draft of my manuscript, thereby prompting major revisions and expansions. Subsequently, Stewart Ethier graciously did the same, with similar results. It's a pleasure to acknowledge their help and support. Additionally, I've benefited from communications with Sergei Maslov, Kim Lee, Steve Jacobs, Michael Canjar, Norm Wattenberger, and Nathaniel Tilton.

Sag Harbor, NY, USA   N. Richard Werthamer

Introduction

Considering the large number of blackjack books published since that original statistics paper more than 60 years ago, it may well be wondered what I could possibly add. After all, the Basic Strategy recipe for playing one's hand has been known and unchallenged for decades. Card counting methods for sizing one's bet, as a complement to Basic Strategy, have also been fully developed. But the many previous publications have shortcomings and missing pieces, which I fill in and clarify. The result is a systematic description of each major aspect of optimal play: how best to play the cards in your hand, how best to assess the odds expected for the next round, and how best to use the odds to adjust the amount you bet on that round.

Furthermore, I emphasize the trade-off between simplicity and performance: I discuss how easy each aspect of optimal play is to actually use in a casino, and I recommend simplified versions, where desirable, which give nearly optimal performance. Any serious player should be carrying out a number of distinct mental processes simultaneously, rapidly, and accurately; he may well wish to drop the least valuable ones, to ensure his fidelity to those more worthwhile—as well as to lower his stress level.

I contribute much new material as well. The most significant is to that third major aspect—how you, as a player, should manage your cash. (I'll usually refer to you in the third person, as Player, with masculine gender assumed, and the dealer, correspondingly, as Dealer, feminine gender.) If you are advised to increase your bet when you've detected favorable odds, then it's likely you'll immediately ask, "By how much"? Closely related is how you can control the risk of a losing streak that wipes out the total capital you brought to the casino. This risk depends critically on the scale of your betting relative to your capital; most players probably lose more money from inadequate capitalization than from non-optimum play.

Almost all previous authors on blackjack either avoid the capital management issue altogether or give vague, hand-waving impressions. Almost no one supplies an answer that seems convincing, let alone thoroughly justified, and the better work is largely in restricted Web chat rooms. In contrast, I demonstrate the optimal relationship between the amount bet on a hand and its indicated odds, a relationship linked to Player's risk tolerance.

Also original is the important concept of Counter Basic Strategy, with the procedure described in Sect. 5.2 and the derivation sketched in Sect. 10.2. And I fully analyze several underappreciated refinements to betting strategy. Two of these, referred to in the literature as back-counting and Kelly betting, are usually mentioned only in passing. A third is betting on multiple simultaneous hands. I show that these can each significantly enhance performance. Their use, in fact, is a key to the success of blackjack teams.

My original contributions as described here were first published in a series of journal articles over the period 2005–2008, cited in the References. The articles may be viewed in full, along with related material, on my website, www.blackjack-science.net.

Part I of Risk and Reward gives a straightforward, self-contained guide to blackjack for a general reader, even a beginning or occasional player. Its narrative is entirely nonmathematical: it recommends how best to play the game and what results to expect.

Chapter 1 reviews the rules of blackjack, along with definitions of the key terms specific to the game. Chapter 2 then covers the simplest and most-used technique for playing a hand, widely known as Basic Strategy, in which Player bases his actions on just the cards in his own hand and the one card visible in Dealer's hand. Limited to only this information, Player should always bet the same amount on each round.

Chapter 3 goes on to describe the more sophisticated group of strategies known as card counting. In these, Player tracks the cards dealt on the previous rounds in order to estimate the odds on the next round and decide how much to bet. A number of variant counting schemes have been proposed and advocated over the years; Chap. 3 compares and evaluates several of the more accurate and currently popular.

Chapter 4 is in many ways the most significant (and original) in the book. Section 4.1 relates the scale of Player's bet size, the rate at which he raises and lowers his bet dependent on the count, and his tolerance for the risk of losing his entire capital. The way in which these factors and others influence whether or not Player has an edge over the casino is carefully explained. Section 4.2 examines a modified betting scheme that accelerates the expected rate of winning. It also points out the advantages of playing multiple hands simultaneously. Section 4.3 shows how Player further benefits from properly choosing the moments to begin betting at a table and to leave it.

Chapter 5 revisits play of the hand, to examine ways in which it interacts with card counting. Section 5.1 considers modifications to the counting technique that best support a play strategy generalized to become count dependent. Section 5.2, in contrast, identifies the best count-independent play strategy, termed Counter Basic Strategy, for a Player who counts cards and systematically varies his bets.

Chapter 6 collects the key points of optimal strategy from the earlier chapters. It summarizes the techniques of play strategy, of card counting, and of bet sizing, and it reviews the complex of factors, including casino countermeasures, determining the amount of money Player might win or lose. It presents the case that casino blackjack for a single Player is better thought of as an entertainment, which can be enjoyed with only moderate skill at relatively tolerable expense, rather than as a consistent moneymaker. A blackjack team, on the other hand, can use techniques not available to an individual that make it a true business, with attractive rates of return. Chapter 6 should be scanned by every reader, whether or not he has dipped into the other chapters.

Part II then details the analysis justifying those recommendations, intended for the subset of readers who wish to understand the game more deeply and/or follow its mathematics. Part II gives detailed, self-contained derivations of the assertions and qualitative descriptions of Part I, with the topics of Chaps. 7–10 corresponding respectively to those of Chaps. 2–5.

The tools used in Chap. 7 are almost entirely algebra and basic probability, so more advanced mathematics should not be needed to follow it. Chapters 8–10, however, bring in the machinery of vector calculus; college training equivalent to a science or engineering degree should be sufficient to understand some of the derivations, while the more sophisticated procedures of the others are detailed in step-by-step appendices. For those readers who desire added support, I recommend Morse and Feshbach (1953), my own guidebook to applied mathematics and even today a leading authority. Among recent but more narrowly focused texts, for multi-variable calculus I suggest Marsden and Anthony (2007), and for matrix algebra I suggest Shores (2007).

Some further comments are in order on the style of the mathematics. First of all, I have avoided the use of theorems, formally stated and rigorously proved as in typical mathematics publications. Rather, I take the approach of most physicists, which is to narrate a derivation assuming, without explicit formal proof, that each step is valid unless there is some evident reason to question it.

Furthermore, at many points in Chaps. 8–10 where a quantity is, strictly speaking, a rational number (ratio of integers), I approach its analysis by approximating it by a decimal, justified because the number of cards shuffled into the pack is large. (Illustration: the fraction 17/52 is a rational, while the decimal 0.327 is a close approximation to it.) The move from the exact discrete math of earlier blackjack analysts, primarily based in probability and statistics, to closely approximated continuum math opens the door to the powerful tools of calculus, and facilitates new results difficult to obtain in the traditional way. As an example, this second edition has added an important analysis of the probability distribution of Player's capital evolving with the hands he plays, illustrated with an informative new figure; these results are virtually unobtainable without the calculus.

Supplementing the text is an index of terms, each citing its first use, and an extensive list of references.

Accessing Interactivity

A special feature of this second edition is that a number of its figures and tables have been made interactive using *Mathematica* and the Wolfram Language, a remarkably powerful software system, and are mounted on the Wolfram Cloud. The caption of each such figure or table gives the URL, www.wolfr.am/BlackjackScience, of the mainpage in the Wolfram Cloud from which its interactive version can be viewed. Instructions for use are given in the note below each figure. Please allow time for each figure to recompute (this process occurs in the Cloud in real time; the images are not pre-calculated). Some figures occupy a 3D space and may be clicked on and then rotated.

I am grateful to Wolfram Research, creators of *Mathematica*, for agreeing to mount my interactive material on the Wolfram Cloud. I am also indebted to the many talented staff members of Wolfram Research for their active help and guidance with *Mathematica,* all along the way. Among them, I particularly want to single out Andre Kuzniarek for his long-term continued support and Jeremy Sykes for his active facilitation with the Wolfram Cloud.

I'm similarly grateful for the dedicated efforts of editorial and production personnel at Springer. Special thanks to Paul Drougas for his encouragement and direct editorial involvement and, with Jennifer Evans, for persistent efforts to meld interactivity from Wolfram with Springer's own publishing endeavors.

Contents

List of Figures

List of Tables

Part I
Strategy

Chapter 1
The Game

1.1 History of Casino Blackjack and Its Analysis

Casino gambling has been booming steadily for decades, throughout the world. In the United States, from a time before the 1930s when casinos were illegal everywhere, they have grown in number and in popularity. Legalized initially in Nevada, they took off after World War II, particularly in Las Vegas and Reno, then around Lake Tahoe. Regulated casinos later mushroomed in Puerto Rico and, beginning in the 1970s, in Atlantic City, NJ. Casinos then sprang up on river boats in parts of the Midwest and South. More recently, American Indian tribes have been using their distinctive legal status to develop casinos on their reservations, especially in Connecticut (the largest anywhere in the world) but also in Illinois, California, Washington and elsewhere. Older casinos in Las Vegas have even been bulldozed and replaced by much larger and grander ones. In almost every instance, each new casino has stimulated another increment in the public's interest. Total casino gaming revenue has risen year after year, into the tens of billions of dollars in the U.S. alone, although evidence is growing that the market is saturating: some older venues are losing share to newer ones. The expansion has been worldwide as well; Macau and Singapore are particularly striking examples of the casino-building boom, and are drawing customers from their previous U.S. destinations.

By far the most popular game in practically every casino is the slot machine. Slots require no skill—in fact almost no thought—and treat the occasional winner to a celebration of sound and light that cannot escape the attention (and envy) of nearby patrons. Engineered with odds that heavily favor the casino (usually 8% or more on each spin of the wheels, several spins per minute), they are rapid and inexorable money transfer generators from their patrons to their operators.

The table games have also expanded in popularity, proportional to the growth of casinos. These games, chiefly craps and blackjack but also roulette and baccarat, vary both in the range of options they afford Player—and hence in the skills required to best take advantage of those options—and in their odds. Roulette is clearly the most straightforward: no strategy decisions are called for, and the odds on each

N. R. Werthamer, *Risk and Reward*, https://doi.org/10.1007/978-3-319-91385-8_1

spin of the wheel are uniformly 5.26% against Player (on tables with both zero and double zero), 2.70% on European tables with just the zero. Craps offers a number of different bets from which Player can choose, each with its own odds, but all favor the casino and none offer strategy decisions. Even the best choice, on the Pass Line, favors the casino by about 1.4%. The casino edge in baccarat, a particular favorite of Asian Players, is only slightly less: about 1.06% when betting on Bank.

Blackjack, on the other hand, requires Player decisions which have a great influence on the expected return. Apart from the added entertainment value of a game of strategy versus a game of mere chance, blackjack with optimal play has the potential not only to reduce the casino advantage virtually to zero but in fact to become favorable to an astute Player. In blackjack (and also poker, alone among casino games), skill creates the possibility, at least in the long run, to systematically win money. Nevertheless, achieving that edge is difficult: not only does Player have to apply concentrated intellectual effort to approach optimal performance, but the casino has several legitimate tools available to counter him.

The published literature about blackjack is considerable. Prior to the mid-'50s the game had not been carefully analyzed. The then-popular books about casino gambling (Scarne (1961) as one example) made assertions that were not much more than guesswork, some in fact quite wrong. A major breakthrough was the work of Baldwin et al. (1956) published in a scholarly statistics journal that for the first time presented a mathematical approach for analyzing blackjack and an optimized playing technique which (with later extensions and correction of their minor errors) has become known as Basic Strategy. Baldwin et al. were also the first to apply computational methods in their analysis, although their "computers" were of the mechanical adding-machine kind. However, Baldwin et al. were explicit, and correct, that Basic Strategy still has a negative expected return: the odds favor the casino for a game with multiple decks.

A few years later, Edward Thorp (1966) took this approach further, by realizing that the expected return for a given hand varies with the composition of the as-yet-undealt cards, and that a careful scan of the cards as they are dealt contains useful indications of that composition. Thorp proposed a specific scanning scheme, or count, which he claimed could identify instances when the odds were actually in favor of Player. Thorp concluded that Player, by increasing his bet when these instances were detected, could achieve a positive overall edge.

Thorp's work, first published in 1962 as a book titled "Beat the Dealer," created a sensation in both the gambling public and the casino industry. Gamblers rushed to apply his methods, while casinos quickly made rule and procedure changes to blunt what they felt was a serious threat to their viability. Some players, based on casino suspicions that they were "counting," were politely but firmly asked to leave. Soon, however, the casinos' panic subsided and the changes were reversed.

Thorp had also written a program for electronic computer that improved on the work of Baldwin, et al. Later, Julian Braun, an IBM staffer, wrote a more extensive and precise routine. The use of electronic computers, even though primitive then relative to today's, gave results within hours rather than months. Over the next decade, Braun refined his methods, supplied results to other blackjack authors, and eventually published his own book (Braun 1980). Braun and others developed a different form of counting scheme than Thorp's, one that was both simpler and more informative. Several variants of this form of counting, particularly one ascribed to Harvey Dubner, have been in wide use ever since. An extensive history of blackjack and developments in its analysis is given by Snyder (2006).

In the several decades since the work of Thorp, Braun, and others, dozens of books about blackjack strategy have appeared, mostly repackaging, extending and popularizing the conclusions of the pioneers. The more quantitative of the subsequent authors have included Peter Griffin and Stanford Wong. Griffin (1999) went beyond Thorp and Braun to analyze many other aspects of blackjack strategy. Wong (1994), on the other hand, undertook an exhaustive study of one favored card-counting method, using a large-scale computer simulation. Wong programmed a computer to "play" millions of hands of blackjack for each of many possible play strategies within that counting method. By comparing outcomes, he identified the most favorable of those strategies and the overall average return to Player. Subsequently, Grosjean (2000) has reported some quite sophisticated analyses of detailed aspects of blackjack and other casino games.

During those years, several Internet "chat-rooms" on blackjack were set up. Those operating like a club, accessed by a password requiring a fee or member recommendation, became especially popular with blackjack experts. Some careful analyses have been contributed to these Web sites and discussed by their participants, although these have not yet been assembled into the sort of coherent and widely available description that a book represents. I've declined to reference derivations posted only to closed sites, and in fact all my results have been obtained independently.

1.2 Rules, Procedures, and Terminology

Blackjack is a table card game. In a casino, the cards are dealt by a casino employee, here called Dealer. The casino and its staff are also referred to collectively as the House. Players, numbering from one to six or seven, occupy seats opposite Dealer and each plays individually against her, not against the others. (In blackjack tournaments, conversely, the contestants play against each other, not the House; tournaments require very different strategies than the casino game.) Any Player may play more than one position simultaneously, as long as the positions are not otherwise occupied.

Each table uses one or more standard 52-card decks, shuffled together. In many casinos, such as those in the Northeast U.S., tables use either six decks or eight. In other casinos, where regulation permits more flexibility, some tables use one or two decks while the rest use either four or six. Dealers at one-deck and two-deck tables hold and deal the cards with their hands. They reshuffle when they wish. At tables with more than two decks, the cards, after shuffling and cutting, have a distinctive reshuffle marker card inserted somewhere halfway or more through, and are placed in a wooden box called a shoe. The cards are dealt, one by one, out of a slot at one end of the shoe. I also use the term shoe for the totality of cards shuffled together. I'll always refer to the cards un-dealt at any moment of time, either in the shoe or held by Dealer, as the pack.

Each card has an associated value. The value of spot cards (except the ace) equals the number of spots; e.g, a three-spot has the value three. Face cards (i.e., jacks, queens and kings) have value 10, just like the ten-spot. I'll refer generically to all 10-value cards as 10s. The ace takes either of two values, one or eleven, at Player's option, although usually the choice is obvious; I'll discuss shortly the criterion for choosing one value or the other. The suit of a card is irrelevant to its value.

In a casino, money for blackjack and the other table games takes the form of chips, in various denominations, redeemed for cash at a cashier's window and purchased either there or at the tables. At the start of every round, Player selects the amount of his bet on that hand and places the corresponding chips on the table in front of him. The amount of his bet (in chip units) can be varied from round to round, as desired, except that the casino establishes a minimum and a maximum bet size for each table.

When all Players have placed their bets, Dealer deals one card to each Player (beginning at her left) followed by one to herself, face up (the upcard); then a second card in turn to each Player and one to herself, face down (the hole card). Typically, Player's cards are dealt face down in hand-dealt games, face up in shoe games. If Player's cards are face down, he may pick them up and hold them; he must not touch face-up cards.

Generally speaking, each Player in turn has the option of receiving additional cards, one at a time, until Player is satisfied with his hand. Player bases those decisions primarily (but not necessarily exclusively, as discussed in Chap. 2) on the total value of all cards he has received in that hand. Player's objective is to simultaneously maximize the total value of his hand and yet avoid the value exceeding 21. Player's judgment on how to balance these conflicting tasks should also make use of his knowledge of Dealer's upcard.

A two-card hand (held by either Player or Dealer) made up of an ace and a 10 is termed a blackjack (sometimes also called a natural). Because the ace is valued in this case at eleven, the hand's total value is 21. If, after dealing the initial two rounds, Dealer's upcard is an ace, her next action is to offer Player a side bet, called insurance, on whether or not her two cards form a blackjack. A Player electing insurance bets an additional amount equal to half his original bet, and is paid off at 2 to 1. Thus if Dealer does prove to have a blackjack, Player without his own blackjack loses his original bet but wins the same amount back through the insurance. In the opposite outcome, Player loses his insurance bet but continues to play his hand against Dealer's.

In most casinos, Dealer with an ace or 10 upcard then peeks, i.e., checks her hole card, to see if it forms a blackjack. If so, Dealer turns it up to show the Players, pays off any insurance bets, and play of that round is over. Any Players who have also drawn a blackjack tie Dealer; those without blackjacks lose their bets. (Significant for Chaps. 2 and 7 is that, for those hands whose play proceeds beyond this point, information has been gained about Dealer's hole card: when combined with the upcard it does not form a blackjack. For example, with an ace up, the hole card is deduced to not be a 10.) Whether or not Dealer peeks, any Player blackjack—should Dealer not also draw a blackjack and thereby tie—wins and is paid off at 3 to 2. For example, if Player bets $20 on a hand and draws a blackjack, while Dealer does not, he gets his bet back plus another $30.

If neither Dealer nor at least one Player have drawn blackjack (or in casinos where Dealer does not peek), play proceeds. The first Player without blackjack now selects one of his several options:

1. He may receive no further cards, termed standing or also sticking.
2. He may double his bet and receive one and only one further card, termed doubling down.
3. If he has a pair of equal-valued cards, he may split them. Each of the two cards becomes a separate hand and Player doubles his bet: he places his original bet on the first hand and adds an equal amount on the second. Each hand wins or loses its bet on its own merits, irrespective of the outcome of the other. Provided the split pair is not aces, Player then plays each hand in turn: after each receives a second card, Player selects from the various play options (including, in many casinos, doubling) except:

   (a) a two-card ace plus 10, although having value 21, is not a blackjack and is not paid 3 to 2;
   (b) if the second card to a split hand is equal in value to the first, again forming a pair, in many casinos that pair can be resplit and option 3 repeats.
   (c) split aces each receive only a second card and no more, and they cannot be doubled or resplit.

4. If the hand is not a pair, or he elects not to split them, Player may receive another card, called hitting (or drawing). At this point his total hand value (i.e., the sum of the values of its component cards, recognizing the ace's dual value) may or may not exceed 21. If the value is not over 21, he may either stand (option 1 again) or receive a further card (option 4 again). If at any point the value exceeds 21 without an ace valued at 11, a condition termed busting or breaking, Player loses, irrespective of the pending outcome of Dealer's hand: she immediately takes Player's bet and puts his cards in a discard tray to await reshuffle.

A hand with an ace valued at 11 is called soft. Conversely, a hand without an ace valued at 11 is called hard. A soft hand with a value exceeding 21 can revalue the ace back to 1 and becomes hard; the hand value is now less than 21 and Player may either stand or hit. Any further aces drawn are valued at 1 and do not push the hand over 21.

When each Player in turn has finished, Dealer turns up her hole card and plays out her hand. Dealer, in contrast to the Players, has no options whatever. She cannot double down or split pairs, and must proceed according to fixed rules: if her hand value initially is 17 or more (with the first of any aces valued at 11) she must stand, whereas if it is 16 or less she must draw cards until the value exceeds 16. If a soft hand comes to exceed 21, the first ace is revalued to 1 and play of her hand continues until its value exceeds 16.

When Dealer has reached either a stand or a bust, the bets are settled. If Dealer has busted, each Player who reached a stand is paid 1 for 1: he gets his bet back plus an equal amount. Each Player who also busted has already lost; he doesn't tie. If Dealer stands, each Player's hand value is compared to that of Dealer: if Player is closer to 21 than Dealer, he wins and is paid 1 for 1; if Player is farther from 21, he loses his bet; and if the values are equal, the two hands tie (sometimes called a push) and Player recovers his bet. Dealer places the used cards in the discard tray and the next round begins.

At the conclusion of the round during which Dealer has encountered the marker, at shoe-dealt tables, she combines all cards (both the discards and those still in the shoe), shuffles them together, cuts (or invites a Player to cut), again inserts the marker card, and reloads the shoe. The fraction of the shoe dealt out before reshuffling is called the penetration. Some casinos use a shuffling machine to speed up the action; it randomizes a second shoe while the first is being dealt, after which the two shoes are switched. The most highly automated machines accept the discards after each round and continuously shuffle them into the cards remaining in the shoe; the penetration is always zero.

In most casinos, Players may join a table, or leave it, at any time; in others, Player may join only during a shuffle. As shown in Sect. 4.3, there are strategy and performance implications in the decision to join or leave a table.

While these rules are generally in effect at many casinos, variants are prevalent. Among the more frequently encountered departures are:

- Dealer is required to hit a soft 17, rather than stand;
- soft hands may not be doubled;
- doubling a split hand after a second card is drawn (usually abbreviated DAS) is not allowed;
- resplits are not allowed;
- Dealer showing an ace or a 10 does not check her hole card for blackjack;
- Player blackjack at single-deck games pays off at 6 to 5 rather than 3 to 2;
- Player can enter a game only during a shuffle.

Also, some casinos offer additional options to Player, the most common of which is called surrender: when Dealer hasn't drawn blackjack, Player may elect to abandon any two-card hand, at the cost of losing half his bet.

Chapter 2
Playing the Hand

2.1 Basic Strategy

Even the most occasional Player wants to avoid egregious mistakes. He's looking forward to some fun, and losing heavily isn't. He would like some guidance on how best to make all those choices he has available and, of course, what the odds are when he does make the best choices.

Every Player has information that is actually quite pertinent to his choices: he knows the identity of the cards in his hand and he knows the value of Dealer's upcard. Even this small amount of data is enough to provide a straightforward but powerful strategy for playing each hand, one that reduces the House advantage to much less than that of any other game in the casino. That recipe for play is usually called Basic Strategy.

Basic Strategy is defined as the class of play decisions with the best odds, given only the values of Dealer's upcard and of Player's hand; the values of his individual cards here serve only to distinguish hard hands from soft and to identify pairs for possible splitting. Basic Strategy also assumes that each split hand, after the second card is dealt to it, is played with the same decisions as for unsplit hands. Optimal Basic Strategy is that Basic Strategy with the best odds for the specific number of decks in the shoe, here labeled D. Discussed later are small improvements in play from use of the individual values of Player's first two cards, as well as that of the second card dealt to a split hand. Chapter 3 lays out the further improvements in optimal play enabled by information on cards dealt in previous rounds.

Optimal Basic Strategy is the same for all numbers of decks between 3 and 6; because of its independence from deck number in this range, it is frequently referred to as Generic Strategy (to follow Wong (1994), pp. 26–27 with DAS, and Vancura and Fuchs (2016), pp. 28–29 without DAS). The guidelines for Generic Strategy are shown in Table 2.1. The final element of Generic Strategy is to always refuse the insurance bet.

The prescriptions of Generic Strategy strongly reflect one overriding characteristic: Dealer's hand is weak when showing less than 7, especially 5 or 6; and is strong

© Springer International Publishing AG, part of Springer Nature 2018
N. R. Werthamer, *Risk and Reward*, https://doi.org/10.1007/978-3-319-91385-8_2

Table 2.1 Generic strategy (Optimal Basic Strategy for 3–6 decks)

Action	Hand value	When Dealer shows	
Double	11		2 through 10
	10		2 through 9
	9, soft 17 and 18		3 through 6
	Soft 15 and 16		4 through 6
	Soft 13 and 14		5 or 6
Split		DAS allowed	DAS not allowed
	Aces and 8 + 8	Any	Any
	2 + 2 and 3 + 3	2 through 7	4 through 7
	4 + 4	5 or 6	
	6 + 6	2 through 6	3 through 6
	7 + 7	2 through 7	2 through 7
	9 + 9	2 through 9, but not 7	2 through 9, but not 7
Stand on	13 and higher		2 or 3
	12 and higher		4 through 6
	17 and higher		Ace, or 7 through 10
	Soft 18 and higher		2,7 or 8
	Soft 18, 3 or more cards		3 through 6
	Soft 19 and higher		9, 10 or ace
Surrender, when available	15		10
	16, but split 8 + 8		9, 10 or ace

when showing more than 6, especially ace or 10. When Dealer is weak, Player should draw conservatively, standing on as little as 12 or 13. When Dealer is strong, Player should draw aggressively, hitting until reaching at least 17. Furthermore, Player should double or split a number of two-card hands when Dealer is weak, while doubling or splitting very little when Dealer is strong.

The indication of strength or weakness from Dealer's upcard reflects, in turn, the fact that she is required to draw to 16 or less. A two-card hand showing a 6 is more likely than not to have a value in the range 13–16, requiring her to draw with a greater than even chance of a bust on the next card. In contrast, a hand showing a 10 (even without a blackjack) is more likely than not to have a value of 17–20, requiring her to stand and posing a strong challenge to Player.

Table 2.2 lists the first round (or reshuffle round) expected return with Optimal Basic Strategy, for four variants of the rules for pair splitting (although both DAS and resplits are allowed in most casinos). The returns, when plotted vs. the inverse number of decks, $1/D$, in Fig. 2.1, are virtually indistinguishable from straight lines. Table 2.2 also lists the slopes of those lines, recurring in Sect. 7.2.2, and the additional expected return from taking advantage of surrender.

As seen in Table 2.2 for the game with pair resplits and DAS, the return for an 8-deck shoe is -0.0043; the reshuffle return improves to $+0.0014$ for a single-deck shoe ($1/D = 1$). Without resplits or DAS, the single-deck return is -0.0001.

Table 2.2 Optimal Basic Strategy expected return ($\times$ 100) vs. number of decks

D	No DAS, no resplit	No DAS, resplit	DAS, no resplit	DAS, resplit	Surrender increment
∞	−0.6902	−0.6510	−0.5704	−0.5094	+0.0932
8	−0.6901	−0.5730	−0.4877	−0.4310	+0.0825
7	−0.5974	−0.5617	−0.4758	−0.4197	+0.0810
6	−0.5839	−0.5468	−0.4599	−0.4047	+0.0790
5	−0.5601	−0.5258	−0.4378	−0.3837	+0.0764
4	−0.5274	−0.4943	−0.4046	−0.3523	+0.0726
3	−0.4731	−0.4420	−0.3495	−0.3000	+0.0662
2	−0.3621	−0.3349	−0.2368	−0.1930	+0.0553
1	−0.0147	+0.0018	+0.1143	+0.1419	+0.0236
Slope	0.336	0.323	0.344	0.325	

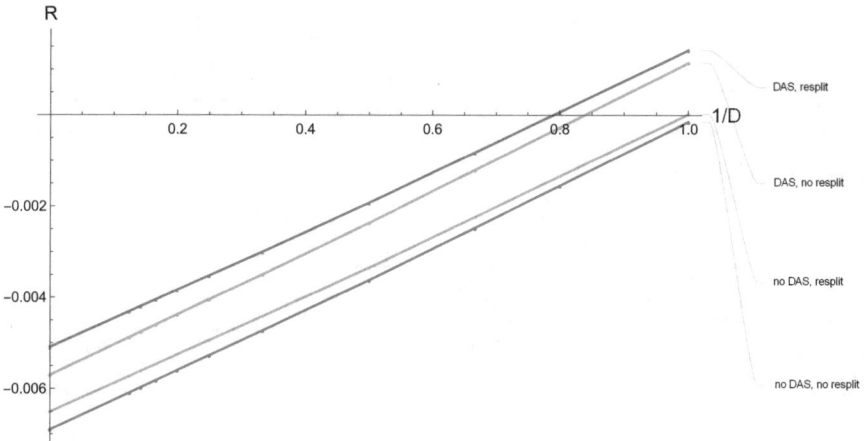

Fig. 2.1 Expected return, reshuffle round, vs. inverse number of decks, four pair splitting rules (the D values from Table 2.2, with 3/2 and 5/4 computed as well)

Thus Optimal Basic Strategy in a one-deck game has nearly even odds, slightly positive or negative depending on the rules for split pairs. For the other size shoes typically encountered, the reshuffle return with resplits and DAS is −0.0040 for six decks and −0.0035 for four decks; without resplits or DAS the respective returns are −0.0058 and −0.0053. The return improvement from surrender, also listed in Table 2.2, decreases with decreasing deck number, from the 0.0008 for eight decks down to just 0.0002 for a single deck.

The specifics of optimal card play also vary with number of decks. Table 2.3 lists the changes in play from Generic Strategy of Table 2.1, for one and two decks. Optimal Strategy for more decks than 6 differs in that $4 + 4$ is not split vs. Dealer's 5; soft 13 is not doubled vs. upcard 5 for more than eight decks; and, for more than 26 decks (!), soft 15 should not be doubled vs. 4 (as first noted by Griffin (1999, p. 176)).

Table 2.3 Differences in Optimal Basic from Generic, one and two decks

Action	Hand value	When Dealer shows			
		D = 2		D = 1	
Double	11	Ace			Ace
	9	2			2
	8				5,6
	Soft 19				6
	Soft 17				2
	Soft 13 and 14				4
Split		DAS	No DAS	DAS	No DAS
	2 + 2				3
	3 + 3			8	
	4 + 4			4	
	6 + 6		2	7	2
	7 + 7	8		8	
Stand on	Soft 18 and higher		Ace		Ace
Surrender, when available	15	Not 10		Not 10	
	16	Not 9		Not 9	

Table 2.4 Generic strategy applied to other numbers of decks

	Reduction in expected return × 100, Generic vs. optimal			
D	No DAS, no resplit	No DAS, resplit	DAS, no resplit	DAS, resplit
∞	0.0008	0.0008	0.0008	0.0008
3 through 6	0	0	0	0
2	0.0026	0.0026	0.0031	0.0031
1	0.0271	0.0272	0.0306	0.0310

The availability of DAS significantly increases the advantage of splitting pairs, enough so that more pairs should optimally be split than without it. Those added (at least in Generic Strategy) are $2 + 2$ and $3 + 3$ against Dealer upcards 2 and 3, $6 + 6$ against upcard 2, and $4 + 4$ against upcards 5 and 6; still more pairs are split for one and two decks. With DAS allowed, Player's expected return increases by about 0.0014 with resplits and 0.0012 without.

Conversely, applying Generic Strategy to one or two decks is only slightly suboptimal: as seen in Table 2.4, the expected return degrades by at most about -0.0003 for one deck, depending on the rules for splitting. These reductions are so small, at least for 2 decks, that they are entirely outweighed by the simplicity of using just a single uniform strategy for any number of decks. With one deck, however, the return improvement from Optimal rather than Generic might be enough to prompt consideration of switching; but a very serious Player may elect instead to adopt either the more complex, composition-dependent (but count-independent) strategy discussed below, and/or a count-dependent strategy as per Sect. 5.1.2.

2.1.1 Expected Return with Variant Rules and Procedures

The rules variation most significant for Player is when Dealer doesn't stand on soft 17, but rather hits soft 17 and stands on soft 18 or more. Player's expected return worsens: with resplits but not DAS, by -0.0022, at least for six and eight decks. The playing strategy remains mostly the same, but with some added situations where doubling is recommended: double 11 against Dealer ace, double soft 18 against Dealer 2, double soft 15 against Dealer 4, and double soft 19 (!) against Dealer 6. These additional doubles arise from Dealer's acquiring a stronger position, so that more aggressive play is required.

The variation where soft hands can't be doubled is more benign; here, the expected return worsens by -0.0008. Not being able to resplit a split hand worsens the expected return by -0.0004.

The variation in which Dealer does not peek (i.e., does not check her hole card for possible blackjack when showing an ace or 10) requires Player to make decisions on whether to double or split without the knowledge that she doesn't have blackjack. As a result, optimal play changes slightly for those upcards: don't double 11 or split 8+8 against Dealer 10, and split no pairs (i.e., don't split aces or 8+8) against Dealer ace. The expected return worsens by -0.0011.

2.1.2 Expected Return vs. Return on Investment

Player's expected return is not the same as his return on investment (ROI). The former is defined as Player's average cash win or loss per hand, per unit base bet. The ROI, in contrast, is defined as the same cash increment but instead per typical amount of cash risked in playing the hand. Since hands that are doubled or split require a total bet of twice the initial bet (or more, in the event of DAS or resplitting), the expected return is clearly larger in magnitude than the ROI. A computation gives the ratio as about 1.12 for Basic Strategy play.

Although this distinction is slight for blackjack, it is highly significant in craps, where the best choice of bet has ROI of about -0.0084. But this is the result of combining an initial unit bet on the Pass Line, having expected return of -0.014, with an equal bet on Free Odds, allowed on an average of 2/3 of all Pass Line bets, having an expected return of zero (i.e., no House advantage in its payoff). This ROI is only a bit more unfavorable for Player than Basic Strategy in blackjack. But Player nonetheless loses an average of 0.014 units per wager, much worse than Basic Strategy.

The concept of ROI will appear again in Sect. 9.1.5, where it plays a significant role in optimal betting.

2.2 Composition-Dependent Play

The focus thus far has been on Basic play strategies, using just the total value of Player's hand and Dealer's upcard. But in fact Player also knows the specific cards in his hand, not just their combined value. A reasonable question is whether, for example, play might be different for a hand whose first two cards are 2 and 6 than for a hand with 3 and 5, even though both have value 8. The more general question is whether the expected return can be further improved through composition-dependent play that takes account of the composition of those first two cards, not just their total value. The improvement, as seen in Table 2.5, is only slight.

Four, six, and eight-deck games have a best play strategy with only a tiny amount of composition dependence. The composition dependence of a two-deck game is greater, but still improves the return only slightly. A one-deck game benefits from playing with a composition-dependent strategy, but the return increase, about 0.00037, is still of interest only to the very serious Player; his return would then range from +0.0002 to +0.0018, depending on the splitting rules. The strategy details are laid out in Sect. 7.3. Basic Strategy, of course, is defined without composition dependence; and composition-*independence* is in fact optimal with a large number of decks, nearly so with as few as four decks.

Table 2.5 Increase in expected return from composition-dependent play vs. Optimal Basic Strategy (resplits, no DAS)

D	∞	8	6	4	2	1
Return increase $\times 100$	0	0.0017	0.0028	0.0052	0.0132	0.0366

Chapter 3
Tracking the Cards

So far, several items of information have been pointed out that help Player make decisions. The total value of the cards in his hand and the value of Dealer's upcard together lead to Basic Strategy. The specific composition of Player's hand is additional information of potential use but, as seen, provides at best only a slight improvement in return.

The most important information not yet used is the values of the cards dealt in previous rounds. If these could be tracked, Player would gain insight into the composition of the remaining pack and whether it's more or less favorable to him than immediately after the shuffle. Exploiting the information contained in the cards dealt during prior rounds is the basis for so-called card counting.

The effect underlying the importance of card counting is that the composition of the pack fluctuates as cards are dealt, and the expected return is sensitive to such fluctuations. Although the *average* likelihood (averaged over many shuffles) of drawing a specific value for the next card is always the same, 1/13 in the instance of a 5, the likelihood in any given pack can vary considerably from its average. If a shuffle has clustered the 5s mostly in the first part of the shoe, so that they are dealt in early rounds, the second part will offer a lower than average likelihood of drawing another. Furthermore, the deviations of the likelihoods from their average will increase as more cards are dealt.

The success of card counting in blackjack stems from the fact that Player's expected return on a given hand varies with the likelihoods for that hand. (This is in contrast to most other card games; in baccarat, for example, the expected return varies negligibly with pack composition, and card counting is essentially useless.) In blackjack it is possible, and not uncommon, for the return to shift from slightly unfavorable at the start of a shoe, as seen in the previous chapter, to significantly favorable later in the pack. A Player who can gauge the shifts in return can then adjust the size of his bet accordingly. In those situations where the return has swung in Player's favor, he can increase his bet. If the favorable periods are sufficiently common and if during them the bet size can be increased enough, Player can develop a net advantage over the House.

© Springer International Publishing AG, part of Springer Nature 2018

N. R. Werthamer, *Risk and Reward*, https://doi.org/10.1007/978-3-319-91385-8_3

There are two important issues for developing a practical card counting method. The first is to decide what indicators of pack composition are meaningful and yet not beyond Player's capabilities. Clearly, if Player could count separately and simultaneously each of the ten different card values as they are dealt, he would have the maximum amount of information to predict the return on the next round. But rapidly updating ten distinct numerical registers in one's head, and arithmetically processing them to arrive at the current expected return, is more than almost any human can handle. Thus the challenge is to find a simpler scheme that still retains as much predictive power as possible. Fortunately, to a very good approximation, only a single combined register is needed. Several variants of such a combination are detailed below, along with the objective measures by which they can be evaluated and compared.

The second issue is to identify a rule for adjusting bets according to the indicated return advantage. Although bets need to be increased during periods of favorable return, the bet on the next round should not jump by a large multiple just because the expected return has moved from slightly negative to barely positive; intuitively that seems to escalate risk unduly. The key to finding an optimal strategy for bet sizing is to balance reward with risk. The next chapter uses this key to develop a class of optimal betting strategies together with the risk each entails.

3.1   Linear Counts

Even if no one is able to count each card value separately and simultaneously, it is quite feasible to track a single linear combination of those numbers. We introduce what we call the counting vector, a set of ten numbers labeled from 1 to 10. We call each number in the set an element, and its label corresponds to a card value. The ace is here always labeled 1 (although some authors consider the ace to follow 10, i.e. 11). Typically the elements of the counting vector are chosen to be integers, most simply to be either plus one, minus one, or zero; in general, though, they could be any positive or negative numbers, not necessarily integers.

Next, we introduce the running count. At each shuffle, the running count resets to zero. If the first card dealt thereafter has value j, the running count increments in the jth element of the vector. If the second card dealt has value k, the running count increments again, now in the kth element. With every succeeding card, the running count further increments by the element of the counting vector corresponding to the value of the card dealt. (An example is shown in the next subsection.) Maintaining a running count throughout the dealing of a shoe, although at first seeming like a difficult task, can be mastered with a moderate amount of practice and concentration; it is, after all, only a single number, not ten of them, and usually an integer at that. A blackjack game program on a personal computer, or other electronic device, is useful for practicing card counting.

The running count may also be recognized as a linear combination of the numbers of cards dealt of each value. Thus, if there actually were an individual who could keep ten separate registers in his head, the best possible use he could make of this information for blackjack would be to form just that one linear combination corresponding to the running count!

Finally, we introduce what most authorities call the true count, or count per deck. The true count is the running count divided by the number of *decks* remaining in the pack, i.e., the number of remaining cards divided by 52. Although estimating this ratio in one's head, on the fly, is obviously an added complication and a nuisance, it isn't that difficult and is imperative for gauging the expected return. Running count by itself is not a quantitative measure of pack favorability, whereas true count is.

Fortunately, the number of decks remaining does not have to be known accurately; a rough estimate is sufficient. Useful is the fact that Player, on average, draws about 2.7 cards in playing his hand while Dealer draws about 2.8; thus a table with just one Player consumes 5.5 cards per average round, or about 9.5 rounds per each 52-card deck in the shoe. Player can simply track the number of rounds played since the last shuffle in order to approximate the number of decks remaining un-dealt. For example, a six-deck shoe contains cards sufficient on average for about 57 rounds, so 19 rounds after a shuffle corresponds to a pack with about four decks remaining. Even more simply, some experienced Players estimate the decks remaining, with adequate accuracy, just from glancing at the height of the discard pile.

3.2 Choosing a Counting Vector

The majority of popular books on blackjack that discuss card counting advocate one or the other of two simple counting vectors, known as Hi-Lo and Hi-Opt and shown in Table 3.1.

As an example of keeping the running count, consider observing the cards 3, 10, 5, and 7 dealt in succession following a shuffle: the running count (with either Hi-Lo or Hi-Opt) would then be $+1 - 1 + 1 + 0 = +1$. Both schemes call for assigning $+1$ to card values 3 through 6, and -1 to value 10; values 7 through 9 are ignored. The schemes differ only in whether or not to pay attention to values 1 (i.e., ace) and 2. Since the Hi-Opt scheme allows Player to ignore more values in keeping his running count, it's clearly simpler than Hi-Lo. The latter, however, turns out to be a more accurate approximation for gauging expected return.

These, and a number of other schemes that have been proposed, can be assessed quantitatively as to merit. There exists an *optimal* counting vector (which we derive

Table 3.1 Most-recommended counting vectors

Vector element	Card value									
	1	2	3	4	5	6	7	8	9	10
Hi-Lo	−1	+1	+1	+1	+1	+1	0	0	0	−1
Hi-Opt	0	0	+1	+1	+1	+1	0	0	0	−1

Table 3.2   Other balanced counting vectors

Vector element	Card value										BC
	1	2	3	4	5	6	7	8	9	10	
Optimum (vb)	−1.28	+0.82	+0.94	+1.21	+1.52	+0.98	+0.57	−0.06	−0.42	−1.07	1.000
Ultimate	−9	+5	+6	+8	+11	+6	+4	0	−3	−7	0.998
Halves I	−1.5	+1	+1	+1	+1.5	+1	+0.5	0	−0.5	−1	0.994
Halves II	−1	+0.5	+1	+1	+1.5	+1	+0.5	0	−0.5	−1	0.992
Revere	−1	+0.5	+1	+1	+1	+1	+0.5	0	0	−1	0.975
Hi-Lo											0.967
Hi-Opt											0.874

later), in the sense of giving a nearly exact measure of the expected return of a round based on the pack composition at its start. The difficulty, though, with using this optimum in actual play is that its elements are not integers; the task of rapidly adding and subtracting a series of decimal numbers together in one's head is too hard. Hence an approximation to the optimum is needed that involves only simple integers, especially just the very simplest, $+1$, -1 and 0.

The appropriate figure of merit for an approximate counting vector has been called its betting correlation by Griffin (1999, p. 44) and here labeled BC. (It is defined precisely in Eq. (8.30), with a slightly different label.) The optimal counting vector has a BC of exactly one, and every approximate counting vector has a BC less than one. The closer BC is to one, the better is the approximation. Hi-Opt has a BC of 0.874, while Hi-Lo has a BC of 0.967. The fact that Hi-Lo's BC is significantly closer to unity evidences its superior accuracy relative to Hi-Opt.

Table 3.2 displays a handful of other approximate counting vectors proposed in the literature, alongside the optimum. (For some of these vectors I've divided by a factor of two all elements as usually quoted, for ease of inter-comparison; the scale change has no effect on performance.) They're listed in inverse order of their BC, showing that slight performance improvements can be achieved at the cost of considerably increased complexity, as decimal numbers are approximated ever more closely by integer or half-integer ones.

In my opinion, Hi-Lo strikes a good balance between simplicity and closeness to the optimum. Either one of the Halves vectors gives an excellent approximation, with complexity only somewhat greater than Hi-Lo and much less than Ultimate. A convenient way to count the 0.5 elements in the Halves (as well as in Revere), popularized by Snyder, is to assign 1 to the red cards, 0 to the black; reversing the colors, of course, works just as well. Similarly, count the 1.5 elements using 2 for one color, 1 for the other. We drop Ultimate, whose tiny performance increment over the Halves hardly warrants the extra effort. An informal survey of Players by Wong in 2003 indicated that Hi-Lo was favored by more than a third of respondents who count, far more than any other method and preferred by 4 to 1 over Hi-Opt. Unbalanced counts (see below) were used by just over a quarter, with the K-O method predominating.

3.3   Unbalanced Counting Vectors

The counting vectors listed above all have the characteristic that they are balanced: the sum of the elements (but weighting the 10th element by a factor of 4, since four distinct card faces have value 10) is constrained to equal zero. As a consequence, the count immediately after a shuffle resets to zero, as stated above. Several authorities (see particularly Snyder 2005, and Vancura and Fuchs 2016) advocate unbalanced counting vectors, in which the constraint is relaxed: the reshuffle count is nonzero, and depends on the number of decks. The more widely used unbalanced counting vectors, along with their betting correlation, are shown in Table 3.3.

But in fact any fixed constant can be added onto each and every element of a balanced counting vector (such as those listed in Tables 3.1 and 3.2), thereby unbalancing it, without changing its betting correlation, BC. In other words, every unbalanced counting vector has a balanced counterpart that performs identically. The main advantage, then, of unbalanced counting vectors is that ones with simple integer elements can be constructed whose balanced counterpart, necessarily with non-integer elements, has a BC superior to the favored all-integer approximation, Hi-Lo.

The disadvantage, though, of an unbalanced vector is that the running count must be adjusted, to subtract off its mean value, prior to converting it to a true count and using it for sizing the next bet. Advocates of unbalanced counts instead recommend using the running count directly, without any conversion to true count. Although this generally forgoes the close link between true count and expected return, the unbalance can be exploited to find that single specific running count value (called the pivot) where it always equals the true count. Then an empirical strategy is suggested for betting and card playing which interpolates between the pivot and the key count, the count at which the expected return is zero.

In my view, unbalanced counts only shift, rather than avoid, the mental arithmetic necessary to accurately assess the current expected return; instead of making the conversion from running count to true count, Player is required to carry out the pivot/key interpolation recipe. And the indicated bet size that results can only be approximate, not optimal. Nevertheless, despite this skepticism, the method and underlying algebra is detailed in Sect. 8.6.

Table 3.3 Unbalanced counting vectors

Vector element	Card value										BC
	1	2	3	4	5	6	7	8	9	10	
K-O	−1	+1	+1	+1	+1	+1	+1	0	0	−1	0.973
Red 7	−1	+1	+1	+1	+1	+1	+0.5	0	0	−1	0.969
Zen	−0.5	+0.5	+1	+1	+1	+1	+0.5	0	0	−1	0.963

3.4   Relating the True Count to the Expected Return

Assuming that Player has selected a counting vector, he must be able to translate the true count, based on that vector, into the current expected return; once the return is indicated, a bet size can be chosen. But, at least for smaller true count magnitudes, the return and the true count are related linearly: the change in return from its initial, or reshuffle, value is roughly proportional to the true count. The choice of counting vector affects only the constant of proportionality: if the scale of a counting vector is modified (for example, a Halves vector using 1, 2 and 3 rather than 0.5, 1 and 1.5), then the constant of proportionality is modified inversely by the same scale factor (in the example, reducing it by a factor of 2).

Figure 3.1 illustrates the linear relationship for several of the more practical counting vectors, all of which cluster quite closely. For each, the expected return crosses over from unfavorable (negative) to favorable (positive) as the true count increases through approximately +1. The coefficient of proportionality between expected return and true count, as well as the cross-over true count, are listed in Table 3.4.

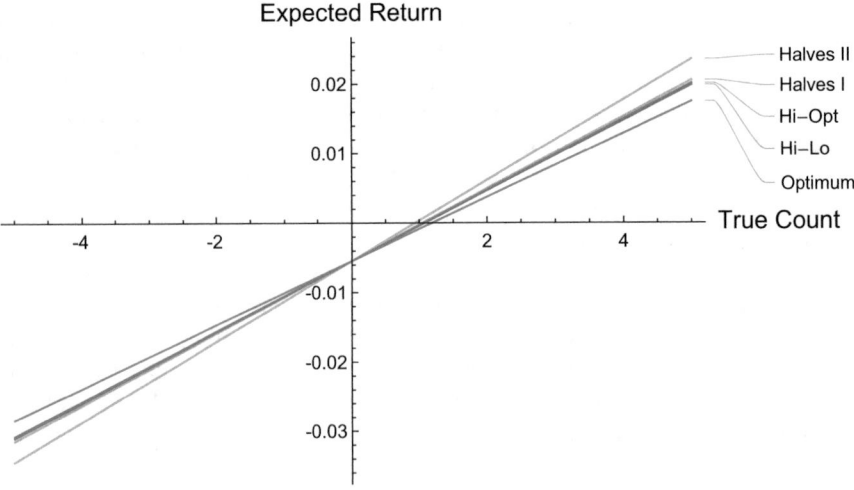

Fig. 3.1 The linear relationship of expected return for small values of true count, for several counting vectors; six decks

Table 3.4 Parameters of expected return vs. true count

	Optimum	Halves I	Halves II	Hi-Lo	Hi-Opt
Coefficient	0.00459	0.00520	0.00580	0.00506	0.00511
Cross-over true count					
Six decks	1.184	1.045	0.936	1.074	1.063
Four decks	1.066	0.941	0.843	0.967	0.957

Remarkably, for all of these counting schemes a true count of about $+1$ is the threshold for a favorable expected return, which increases by about another 0.005 for each further increment in true count by $+1$. In other words, a useful model for expected return in multi-deck games is $0.005 \times$ (true count -1), independent of the counting scheme used to generate the count. For example, a true count of $+5$ indicates an expected return of about $+0.02$. For single-deck games, where the cross-over is roughly at zero, the expected return is instead just $0.005 \times$ true count.

The linearity of return with count holds well for smaller magnitudes of count, within the range between about ± 6. The deviation from linearity becomes significant for values outside this range, as illustrated in Fig. 8.2.

For some of the computations to be reported later, it suffices to model multi-deck blackjack simply as a game with six decks, a reshuffle return of -0.005 and variance of 1.26, penetration of 0.8, crossover true count of $+1$, and a linear relationship between expected return and true count—independent of such rule details as the treatment of split pairs or of the particular play strategy used. This model is similar to the "Benchmark game" of Wong (1994, p. 18).

Chapter 4
Betting

4.1 Yield, Risk, and Optimal Bet Strategies

The crux of Chap. 3 is that the true count is a valuable clue to the expected return on the next round. But Player is concerned operationally only with how much to bet, not with the expected return: he really needs a rule for converting true count directly into bet size. Furthermore, the mental exertion of counting cards and adjusting bet size is only justified to the extent that it gives Player an overall advantage over the casino. An appropriate quantity by which to measure that advantage is the average, over all rounds between shuffles, of the product of the expected return on each round and the bet size on that round, weighted by its probability of occurrence; this is Player's yield, or his average cash flow per round. The average over the shoe is necessary because the first (and probably several) rounds after a shuffle have a negative expected return (except for the single-deck game, as seen in Sect. 2.1). Only when enough of the shoe is dealt to permit sufficiently likely fluctuations in composition that result in a positive return, and the bet is increased correspondingly, can Player make up for his expected losses early in the shoe and achieve a positive yield.

However, yield as defined above depends on the currency denomination of the bet, for example its size in dollars. More useful would be a denomination-independent quantity, the ratio of yield to a measure of bet size. One such measure is the base bet, the bet Player makes on the first round after a shuffle and on other rounds with a negative expected return. But since bets vary from round to round and are sometimes multiplied because of doubling or pair splitting, another measure is the typical bet (defined more precisely in Sect. 9.1.2), averaged over the shoe; the typical bet can be several times the base bet. For now, adopting the former option, we define the yield ratio as the yield per unit base bet. However, the yield per *typical* bet (the same as the return on investment, or ROI, discussed in Sect. 2.1.2) can be another useful quantity, as we'll see below.

Intimately tied to the issue of bet size is that of risk. A Player who makes very large bets relative to his current cash (his capital, sometimes called his wealth or fortune), especially at rounds where his edge over the House is only slight, obviously

N. R. Werthamer, *Risk and Reward*, https://doi.org/10.1007/978-3-319-91385-8_4

has an increased risk of substantial losses—even of losing all his money. On the other hand, a Player who keeps his bets very small has little or no likelihood of gaining a positive overall advantage. The overriding determinants in specifying a bet size are Player's capital as he starts his session, usually called his trip capital or stake (sometimes his bankroll), and his tolerance for the risk of losing it all, a distressing event traditionally called ruin.

An occurrence of ruin doesn't usually imply that Player loses his entire net worth and can never gamble again. Instead, the stake he's lost is probably only a fraction of those assets; he is then able to tap into the remainder and return on a later occasion with a fresh stake. Over a single playing session without ruin, Player's expected capital would grow at a rate per round given by his yield; but when averaged over all sessions including those that end in ruin, the long-term expected capital growth rate is reduced. That latter rate is here called the effective yield, and is lower than its no-ruin value by a yield reduction factor, conveniently abbreviated as YRF. Plausibly, the YRF and the risk of ruin are complementary: the YRF behaves qualitatively (to within a factor of two or less) like the probability of survival, i.e., one minus the probability of ruin.

In sum, an optimal bet strategy strikes the best balance of risk vs. reward, specifically by maximizing Player's effective yield while fixing his ruin probability at a predetermined value. This criterion leads to a family of betting strategies, depending on the risk value selected.

Several consequences follow from the criterion. The first and most significant is that the optimal bet is sized in direct proportion to (i.e., linearly with) its expected return. But the bet obviously can't be less than zero, even if the expected return is negative; and a very small bet violates the casino rule that it can't be less than the table minimum. Rather, Player makes the base bet on negative return rounds; it may, though, be more than the table minimum. (Discussed later, in Sect. 4.3, is the technique of table-hopping, which avoids betting on most rounds with negative expected return.) Also, bets can't exceed the maximum set by the casino for that table; but Player may wish to cap his bet size below that maximum, at a multiple of his base bet called the spread. The optimal strategy, constrained in this way, is illustrated in Fig. 4.1, a relationship here called a ramp.

Another consequence of the optimization criterion is that the constant of proportionality between bet size (in units of Player's capital) and expected return (a constant we'll call the ramp's steepness) is tied to Player's risk of ruin. A steep ramp, as noted above, forces Player into large bets even with only slightly favorable odds, and hence magnifies the chance of ruin during a losing streak. A shallow ramp, on the other hand, although less risky, dictates small bets even with distinctly favorable odds and so hinders or even prevents Player's expected capital from growing.

The ramp in Fig. 4.1 (blue line) further violates the casino rule that bet amounts must be in discrete multiples of the table minimum, rather than smooth like the ramp. To accommodate that added reality while still approximating the ideal ramp as closely as possible, bets should instead be sized according to a jagged staircase in true count, as schematized in Fig. 4.1 (red line). Since a staircased bet has perfor-

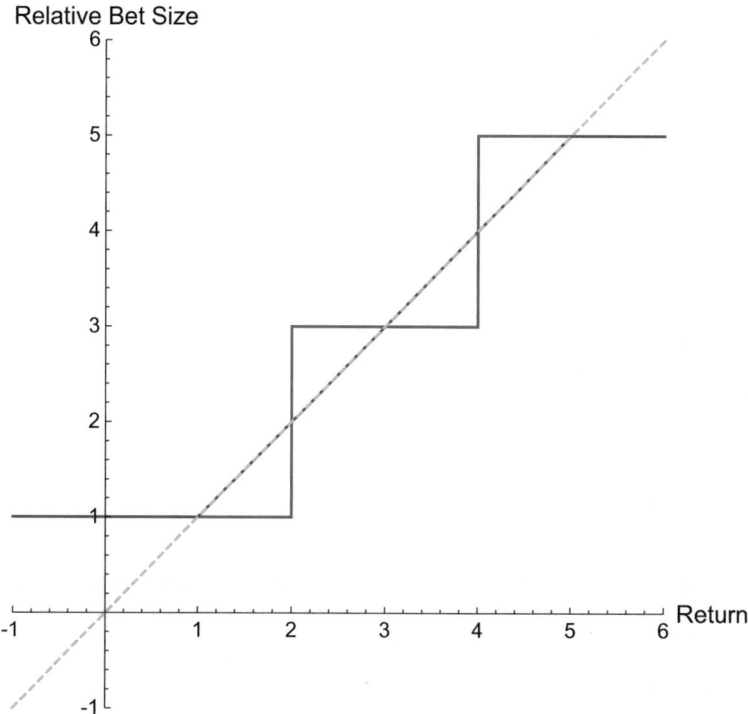

Fig. 4.1 Schematic representation of optimal betting strategy: ramp (blue line: the green dashed diagonal constrained by lower and upper bet sizes of 1 and 5, respectively) and a two-step staircase approximation (red)

mance only modestly reduced from that of the corresponding ramp, as investigated in Sect. 9.1.6, most of the discussion here still focuses on the idealized ramp.

Although Player would naturally like to achieve simultaneously the goals of both high yield and low risk, the two instead are mutually exclusive and require a tradeoff. To pare down the multiplicity of variables and factors influencing the tradeoff, concentrate for now on just two representative examples of Player. One typifies a dedicated, intensive professional (call him the Lifetimer), spanning his playing career; the other, the Weekender, typifies a more casual player, although still a card counter. Assign the Lifetimer a million rounds (about 10,000 h at a typical 100 rounds per hour) and trip capital of a thousand base bets; assign the Weekender a thousand rounds (about 10 h) and trip capital of a hundred base bets. Assume for each the model Benchmark game, along with a bet spread of 10.

Computations for these examples find qualitative similarities. Risk (blue) plotted vs. ramp steepness, in Fig. 4.2 for the Lifetimer, shows a sharp minimum for a small steepness value and a rise for larger ones. Effective yield (red) also rises with steepness, but with a shallow maximum at a larger value.

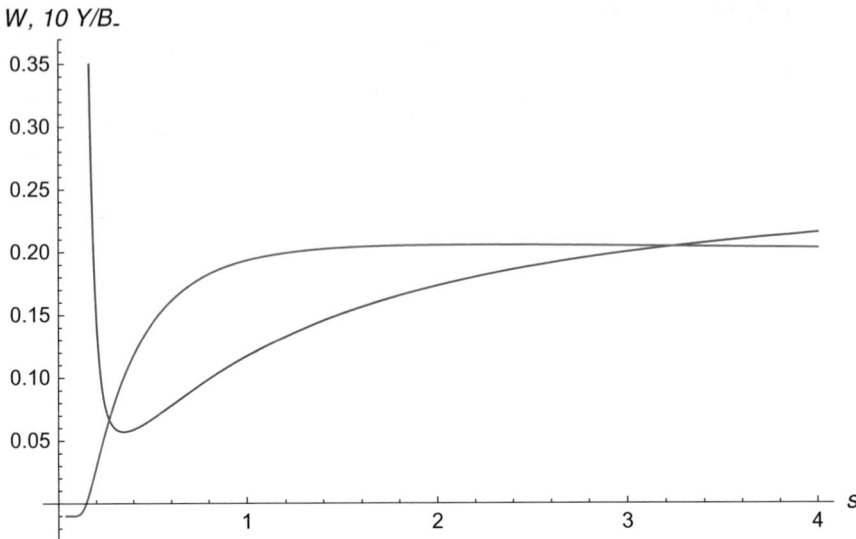

Fig. 4.2 Risk of ruin (W, curve with a minimum, blue) and effective yield ratio (Y/B_-, curve with a maximum, red) vs. ramp steepness (labeled s). Plotted for the Lifetimer example: bet spread $=$ 10, capital $= 1000$ base bets, rounds $= 1,000,000$, 4 decks, penetration $= 0.8$, no DAS, no resplits. To interactively select other values for these parameters, browse to the Wolfram Cloud website www.wolfr.am/BlackjackScience and select the appropriate value. This figure is copyright by N. Richard Werthamer and used by permission

Thus, effective yield increases with increasing risk, as plotted in Fig. 4.3, but only between the points where risk is minimum (at a low steepness) and where effective yield is maximum (at a high steepness). All points in between are legitimately Optimal, according to the criterion of maximal effective yield for a given risk; and Player must then address his yield vs. risk tradeoff.

The Lifetimer, since he's very well capitalized, might at first choose the point that maximizes his effective yield, irrespective of risk; while the Weekender, under-capitalized, might instead choose to minimize his risk at a sacrifice of effective yield. But neither endpoint on the yield vs. risk curve is a sensible strategy: moving a bit away from either endpoint provides a much better yield/risk tradeoff. Instead, intermediate points along the curve seem more reasonable targets for bet strategy; we'll take two particular ones for closer study.

Call one of these the minmax (or MM) point. It confronts the yield/risk tradeoff directly by maximizing the effective yield *per unit* of risk. Call the other the HJY point, originally identified by Harris (1997). Harris, in particular, derived this condition by maximizing Player's ROI for a given spread. For now, the two points are marked on the curves in Fig. 4.3, and Table 4.1 lists the corresponding operating parameters: steepness of the bet ramp, risk of ruin, and effective yield.

For the Lifetimer, the HJY bet point seems to be a reasonable choice: its effective yield is near the maximum attainable, yet with noticeably lower risk of

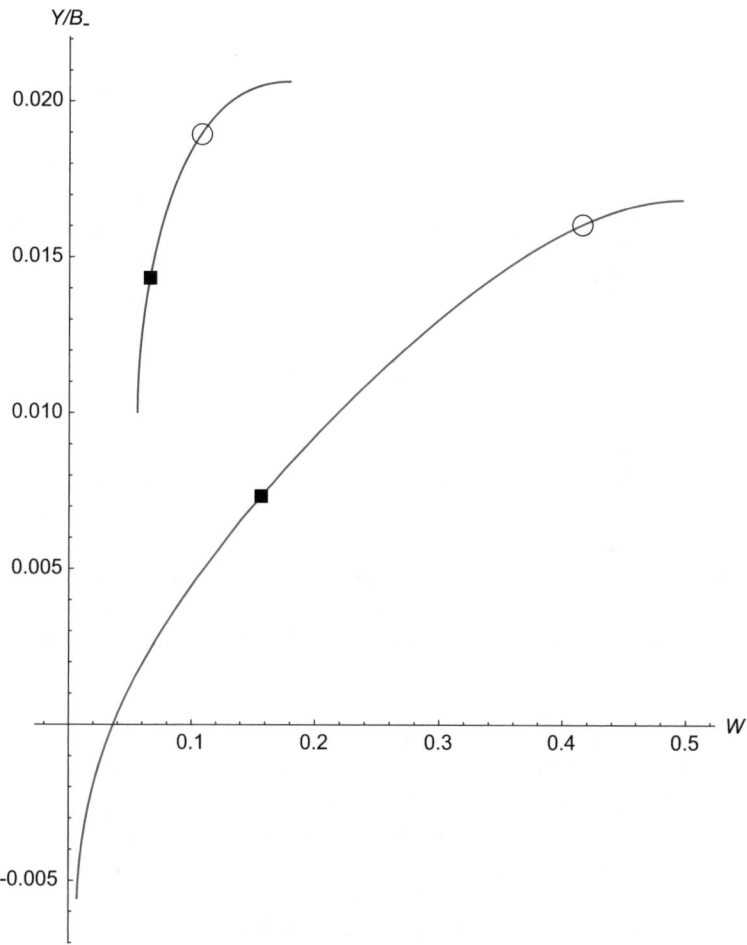

Fig. 4.3 Effective yield (Y/B_-) vs. risk of ruin (W), each with points MM (solid squares) and HJY (open circles). Plotted for both the Lifetimer (blue as in Fig. 4.2: capital = 1000 base bets, rounds = 1,000,000) and the Weekender (red: capital = 100 base bets, rounds = 1000); both curves for bet spread = 10, 4 decks, penetration = 0.8, no DAS, no resplits. To interactively select other values for these parameters, browse to the Wolfram Cloud website www.wolfr. am/BlackjackScience and select the appropriate value. This figure is copyright by N. Richard Werthamer and used by permission

ruin. For the Weekender, on the other hand, the MM point seems more reasonable: its risk, although above the minimum allowed, has a dramatically increased—and positive—effective yield. The HJY point seems too aggressive and dangerous for the Weekender; the MM point seems too timid and unrewarding for the Lifetimer. Although any Player can choose another point from the continuum of optimal curves of Fig. 4.3, the Weekender should be careful to avoid too shallow a ramp: at less than half of the MM steepness the yield is negative!

Table 4.1 Operating parameters for the MM and HJY betting points, Lifetimer and Weekender examples

Player	Point	Steepness	Risk of ruin	Effective yield ratio
Lifetimer	MM	0.52	0.074	+0.013
	HJY	0.94	0.118	+0.017
	Max	2.22	0.193	+0.019
Weekender	Min	0.12	0.007	−0.005
	MM	2.80	0.136	+0.006
	HJY	9.37	0.409	+0.015

4.1.1 Factors Influencing Performance

Generalizing beyond these two Player examples, risk and effective yield are determined by several factors. One is his trip capital. A Player with a large stake relative to his typical bet, call it his coverage, can easily afford to raise his bets when signaled by an above-threshold true count, while also riding out most losing streaks and averting ruin. A Player with low coverage, on the other hand, is vulnerable even to a relatively brief (and hence frequently encountered) losing streak.

A second factor is the number of rounds he plays within which he hopes to avoid ruin: the more rounds, the greater the risk.

A third factor is the penetration, the fraction of all cards that have been dealt when Dealer reshuffles. Immediately following a reshuffle, the pack's composition and expected return are known. The bet ramp must overcome a reshuffle return that, although dependent on the number of decks as seen in Sect. 2.1, is negative for multi-deck shoes. But as additional rounds are dealt and the pack's depth increases, its composition can increasingly depart from that at the reshuffle. The distribution of returns on succeeding rounds also varies from the reshuffle value: positive expected returns are encountered with increasing likelihood. The less the penetration, the less the overall proportion of rounds with positive expected return and opportunity for increased bet size, and so the less the capital growth rate for a bet ramp of given steepness.

Table 4.2 shows the operating parameters for the examples of Lifetimer (at the HJY point) and Weekender (at the MM point) vs. penetration, confirming the performance degradation as the penetration decreases. Note that the effective yield is always positive, even for relatively low penetration; the bet ramp steepens to achieve this. But at high steepness the risk becomes quite large, signaling Player to compensate by increasing his coverage.

Yet another factor is bet spread. Virtually every blackjack authority advises that casinos try to protect against card counters by closely scrutinizing Player behavior and requiring such Players when detected to leave the blackjack tables, or possibly even the casino itself—permanently. (One recent guide to the legality of casino actions vis-a-vis Player is Nersesian (2016).) To reduce the likelihood of detection, most authorities recommend that Player apply caution to his bet strategy: they advise restricting bets to no more than a moderate spread over the base bet, typically around

Table 4.2 Operating parameters vs. penetration, Lifetimer (HJY point) and Weekender (MM point) examples

Player	Penetration	Steepness	Risk of ruin	Effective yield ratio
Lifetimer	0.8	0.94	0.118	+0.017
	0.7	1.28	0.209	+0.012
	0.6	1.75	0.319	+0.008
	0.5	2.53	0.449	+0.005
Weekender	0.8	2.80	0.136	+0.006
	0.7	4.70	0.223	+0.007
	0.6	7.90	0.308	+0.006
	0.5	13.28	0.374	+0.004

Table 4.3 Operating parameters vs. spread, Lifetimer and Weekender examples

Player	Spread	Steepness	Risk of ruin	Effective yield ratio
Lifetimer	20	1.20	0.188	+0.032
	10	0.94	0.118	+0.017
	5	1.30	0.143	+0.006
Weekender	20	2.82	0.147	+0.007
	10	2.80	0.136	+0.006
	5	3.17	0.104	+0.004

5 to 10; or, alternatively, restricting staircases to only a small number of steps (4 or 5, for example); the table maximum, by comparison, might be 50 or more times the minimum. Some authorities also suggest not increasing the bet, even if so indicated by the true count, unless the previous hand was won; and, conversely, not decreasing the bet unless the previous hand was lost. Tactics of these kinds (usually called camouflage) all lower the yield and increase the risk. Yet they depend on Player's assessment of the casino's oversight and enforcement practices, and of the balance he wishes to strike between aggressiveness and discretion. Such decisions are primarily judgmental, not amenable to optimization.

The degradation in performance from lowering the bet spread, however, can be computed. Table 4.3 shows this for the HJY Lifetimer and the MM Weekender. Dropping the spread much below 10 sacrifices a great deal of yield advantage.

A final factor, qualitative but intertwined with the others, is Player's risk tolerance—his willingness to end his session with empty pockets.

Since Player's financial performance is influenced by these multiple factors, a recipe for how to maximize that performance, yet still keep risk under control, is not easy to describe. The complexities of analyzing those factors, including the ways in which they interact with each other, is deferred to Part II, Chap. 9. Perhaps the most important of them, though, is Player's capitalization, particularly as measured by his coverage ratio. To forestall ruin for a given number of rounds, Player's coverage must be substantial. For example, a capitalization of 20 base bets (e.g., a $200 stake at a $10 minimum table) typically is barely adequate to avoid ruin for 400 rounds, roughly one evening's session at about 100 rounds per hour. A capitalization of 50 to 100 bets is not too large for an avid player who might spend a dozen hours at the table over the course of several days (like the Weekender example). This reality of blackjack puts extra pressure on Player: in order to limit risk and capture the reward promised by a counting strategy, he must be prepared to bring a large stake to the game. Conversely, under-capitalization can sabotage the performance of an otherwise optimized strategy (as well as prematurely end the gaming trip!).

A Player who begins his session with a certain amount of capital will find that it fluctuates as his session continues. The deviations from its initial value exhibit a probability distribution of increasing width; and the mean of the distribution shifts, increasing or decreasing depending on whether the typical return per hand is positive or negative. Furthermore, as discussed above, Player can experience ruin. It is instructive to examine, at least cursorily, a plot of such a distribution. Figure 9.1 illustrates these features as the number of hands increases; but since details of the plot are complex, it is placed in Part II together with its mathematical basis.

4.1.2 Bet Size in Relation to True Count

As emphasized at the start of this chapter, Player needs a specific and readily implemented recipe for how much to bet on the next hand based on the current true count. The discussion thus far, however, has focused entirely on the connection of bet size to expected return, with an optimal relationship of strict linear proportionality. But as described in Sect. 3.4, the dependence of the expected return, at least for small to moderate counts, is a similarly linear relation. Thus in this count regime the bet size is also in linear proportion to the true count. The constant of proportionality, in turn, depends on the multiple factors within Player's control as listed above: tolerance for risk of ruin, number of rounds within which ruin is to be avoided, capitalization, and bet spread, as well as such game conditions as the number of decks and the penetration at reshuffle.

Much of the computational results above, Tables 4.1, 4.2, 4.3, are confined to the two representative examples of the Weekender and the Lifetimer. Although these examples illustrate the expected performance for typical classes of Player, other conditions may also be relevant. Here we make available an interactive Table to specify alternative parameter values and to read out the results for each of them. Parameters that can be individually selected are: player's ratio of trip capital to base bet, his bet spread and the number of rounds he plays, the number of decks, and the penetration to which they are dealt before reshuffle. For each choice of these parameters, the Table outputs the minimum and maximum allowable ramp steepness, as well as at the MM and HJY operating points lying between the min and max values. Player's choice of steepness can then be selected, within the min to max range shown. Primary outputs are the risk of ruin and the corresponding yield ratio. To work with this Table, browse to the Wolfram Cloud website www.wolfr.am/BlackjackScience.

In this interactive Table, Reader is invited to select a bet ramp steepness, within a given range, from which the risk and yield ratio are displayed. A different

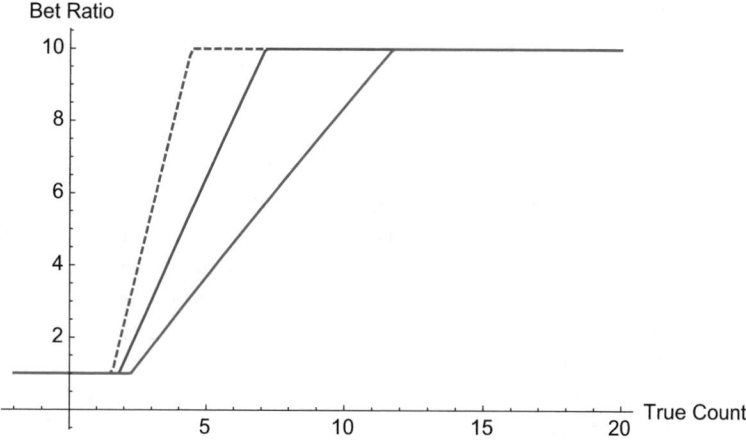

Fig. 4.4 Bet ramp, parameters as in the previous figures, plotted for the Weekender (red) with risk at his MM point and the Lifetimer (blue, solid) with risk also at his MM point. The ramp for the aggressive Lifetimer's HJY point (blue, dashed) is so steep as to be considered impractical. To interactively select other values for these parameters, browse to the Wolfram Cloud website www. wolfr.am/BlackjackScience and select the appropriate value. This figure is copyright by N. Richard Werthamer and used by permission

perspective is to instead invite a Reader's choice of risk, within a given range, with the steepness and effective yield as outputs. Figure 4.4 illustrates this for the same choice of parameters as in Figs. 4.1, 4.2, 4.3; it displays the resulting bet ramp vs true count, avoiding the need to work it out based on the steepness.

4.1.3 Other Criteria

In the discussion above, the yield is scaled to Player's base bet. But another useful ratio, called coverage, is yield relative to his typical bet. As introduced in Sect. 2.1.2, this ratio is Player's return on investment, or ROI. Obviously, the greater the ROI, the less the risk, when other factors are the same. Maximizing the ROI for fixed spread is equivalent to the bet optimality criterion used by Harris (1997) to obtain the HJY operating condition.

4.2 Betting Proportional to Current Capital

One of the earliest advances in blackjack analysis, contemporaneous with the Basic Strategy computation of Baldwin et al. (1956), was an insight by Kelly (1956) applicable to betting. He pointed out the benefits of bet sizes that are always proportional to Player's *current* capital. An underlying rationale is that a Player presently with a few hundred dollars should be betting less than if he had many thousands, irrespective of the capital he started with. Or alternatively, a Player who happens to be winning (or losing) big becomes inclined, quite rationally, to rescale his bet sizes upward (or downward) because his risk has changed.

This concept is mentioned, cursorily, in much of the subsequent blackjack literature. It is typically stated, following Kelly, that with this betting approach the expected capital increases exponentially with the number of rounds, rather than linearly as in the previous section, and that Player notionally can never be ruined. But he avoids ruin only by decreasing his bet size when/as his capital diminishes; he thereby comes to violate the casino's table minimum rule. As a consequence, Kelly betting has been widely regarded as an impractical curiosity (in gambling, but decidedly not in financial investing—see Poundstone (2005) for a popular account, while MacLean et al. (2010) is comprehensive and technical).

What seems like a fatal flaw, though, stems from a misinterpretation of the ruin concept. If Player is indeed experiencing a losing streak, with shrinking capital, and reaches the situation where his next bet, as dictated by the Kelly fraction of that capital, would drop below the table minimum, then his betting scheme dictates that he stop playing even though he still has some capital left. This event, rather than zero capital, is the more appropriate definition of ruin.

With this modified concept of ruin, Kelly betting becomes very much like the approach of the previous section. In both cases the optimal bet size turns out to be directly proportional to the expected return (when positive) on the next round, although the factor of proportionality differs somewhat. Previously, the factor was adjusted to control risk and is proportional to Player's initial trip capital; here instead, the proportionality factor is just his *current* capital. Kelly betting is like risk-controlled betting where Player's capital has increased to the point that he can rescale his bets upward in keeping with his newly-won wealth—or downward during losing streaks.

Rescaling of bet sizes can be frequent when Player is so well capitalized that he is betting large multiples of the table minimum and the bet sizes he is permitted are relatively close together. Alternatively, bets can be rescaled occasionally when Player's current capital has diverged significantly from its starting point. Either way, Kelly betting is a natural extension of risk-controlled betting.

Another frequent criticism of Kelly betting is that the fluctuations of capital that Player experiences (i.e., his variance) are very large; such criticisms typically advocate reducing the fraction of current capital bet per round, from the Kelly value to as little as a half or even a quarter of that. This does reduce the variance, but at the cost of a comparable reduction in the growth rate of mean capital. Such criticisms usually overlook the ruin concept just discussed, whereby Player brings in fresh capital rather than continuing with a very small stake. Another technique for damping variance is to play multiple simultaneous hands, as discussed in the next section.

4.3 Multiple Hands

Playing multiple hands simultaneously is usually permitted as long as there are unoccupied seats at the table, although in some casinos only just after a reshuffle. Taking advantage of this option, and distributing the bet amount among the hands, is an under-appreciated technique for combating large fluctuations (high variance) in capital. It can be shown that the optimal such allocation is to bet an equal amount on each hand; and that, as with betting on a single hand, the total amount should again be directly proportional to the expected return. Although the allocation among multiple hands has the same expected return as placing that same total bet on just a single hand, the variance of the outcome is lower. Furthermore, the reduced variance implies that the risk of ruin is similarly reduced.

This effect is illustrated and quantified in the same interactive Table 4.1, 4.2, 4.3, 4.4 presented above, but which contains an additional parameter for the number of seats played simultaneously; for the previous discussion, the number of seats was set to one. Note that a larger number of seats also encourage adjusting the bet steepness for further improvement in risk and/or yield, as desired.

Kelly betting is similarly enhanced by playing multiple seats simultaneously. Also, the Kelly bettor gains additional advantage from reallocating his capital between the hands. In this technique, Player begins by dividing his capital evenly among the seats and Kelly-betting each hand from its own capital. Following a round (or at least after several rounds) he again pools his capital and redivides it equally among the seats; this permits winning seats to subsidize losing ones. Capital reallocation can measurably increase the growth rate of his median capital and/or reduce its variance.

Several reallocation methods are possible: in one, Player optimally allots for betting on a round a larger proportion of his total capital than he would with just a single seat, by a factor listed in Table 4.4, and distributes that allotment equally among the seats. (As an example, $30 on each of two seats rather than $40 on just one.) As a result, the median capital growth rate is greater by that same factor in Table 4.4, typically between 1.5 and 2.

The factor is less than the number of seats, limited by the correlation between the hands resulting from them all challenging the same Dealer hand. The increased capital growth rate occurs because those seats which are losing, and so might not recover enough to contribute significantly to Player's end-of-session capital, instead are regularly reinforced by the winning seats and restored to a more normal "productivity." While the capital growth rate is increased in this way, the variance of capital is similarly—and undesirably—increased.

Table 4.4 Factor of bet proportion and of median capital growth rate, vs. number of seats

Seats	1	2	3	4	5	6	7
Factor	1.00	1.46	1.72	1.89	2.01	2.09	2.16

This concern is removed with a second method: the bet allotment per round is still distributed evenly among multiple seats, but its proportion of total capital is instead just the same as if it were all bet at just a single seat. (As an example, $20 on each of two seats rather than the $40 on just one.) The median rate of capital growth is less than that of the previous method but is still greater than if the bet had been concentrated at one seat. The advantage here is that the capital variance is correspondingly reduced.

These results are much more pronounced if each of the seats could be played at a separate table, rather than all at the same table: since the tables (and their Dealers) are now uncorrelated, the growth rate of median capital scales directly with the number of seats/tables. Although obviously impossible for a single Player, playing multiple tables simultaneously is easily realized by a team. The team only needs an arrangement to pool and redistribute their capital at appropriate intervals; it need not be nearly as frequent as every round to derive most of the benefit available. Such a team also gains from the size of its capital pool, thereby assuring it a much more steadily expanding stream of winnings than an individual could by himself.

4.4   Back-Counting and Table-Hopping

Performance can be improved even further by a playing technique commonly called back-counting (or sometimes wonging, after its advocate, Stanford Wong). The back-counter begins by counting a freshly shuffled shoe while standing near the table and not betting. When and if the true count crosses a certain positive entry threshold, he sits down and starts betting.

He may, in addition, choose to leave the table when and if the true count crosses certain negative thresholds, either prior to entry (called departure) or subsequent to entry but before the next reshuffle (called exit). If he does leave before the reshuffle, most likely he will immediately turn his attention to another table and repeat the process. If he's playing at a second table before the first is reshuffled, we'll follow Wong and call him a table-hopper.

The optimal table-hopping strategy is specified by three true count thresholds: for entry, for exit, and for departure. Results are shown in Table 4.5, for the two Player examples of the Lifetimer and the Weekender, each at the Wong "Benchmark" game; for computational simplicity, risk is neglected.

The first rows give the yield ratio in the absence of any back-counting, i.e. play-all. The second rows give the optimal true count for entry, without any exit or departure. The third and fourth rows give the jointly optimal entry and exit true counts, without and with table-hopping, respectively. The fifth rows give the optimal exit true count for a table-hopper who enters at a shuffle, without any back-counting. The sixth rows give the jointly optimal entry and departure true counts when table-hopping. The last rows combine all table-hopping procedures into a single jointly optimal set of entry, exit and departure true counts.

Table 4.5 Optimal true count (tc) thresholds and yield ratio for table-hopping

	# tables	Entry tc	Departure tc	Exit tc	Yield ratio
Lifetimer					
Play-all	1	0	$-\infty$	$-\infty$	0.0180
Entry only	1	2.09	$-\infty$	$-\infty$	0.0238
Entry and exit	1	1.89	$-\infty$	−3.57	0.0242
Entry and exit	2	1.81	$-\infty$	−3.18	0.0243
Exit only	2	0	$-\infty$	−1.91	0.0224
Entry and departure	2	2.08	−3.49	$-\infty$	0.0240
Entry, exit and departure	2	1.85	−3.29	−3.45	0.0245
Weekender					
Play-all	1	0	$-\infty$	$-\infty$	0.0052
Entry only	1	2.60	$-\infty$	$-\infty$	0.0113
Entry and exit	1	2.21	$-\infty$	−2.72	0.0116
Entry and exit	2	2.15	$-\infty$	−2.57	0.0117
Exit only	2	0	$-\infty$	−1.54	0.0095
Entry and departure	2	2.58	−5.03	$-\infty$	0.0113
Entry, exit and departure	2	2.14	−5.13	−2.61	0.0117

The Lifetimer example is an aggressive one, such that his play-all yield ratio, without any back-counting, is already an attractive +0.018. Entry, at an optimal true count of +2.1, adds another 0.006 to the yield ratio, so is a valuable addition to his repertoire. Exit after entry adds still another 0.00045 to his yield ratio; the optimal exit true count is around −3.6, while the optimal entry shifts slightly downward to around +1.9. Departure is also of some value: the improvement in yield ratio from departure, even without exit, is nearly equal to that from exit without departure; and the combination of both exit and departure (always, of course, with entry and subsequent table-hopping) gives a yield ratio improvement about double that from either alone.

Similar results also hold for the Weekender, especially considering that the Weekender's play-all yield ratio is so much lower than the Lifetimer's. The yield increments from table-hopping maneuvers are roughly the same for both Players.

Remarkably, the table-hopper who merely exits at a true count of roughly −1.5 to −1.9, and always begins play at a table immediately following a shuffle, accomplishes much of what entry and exit together can. This technique could be particularly difficult for casino surveillance to spot and so provides the table-hopper excellent camouflage with a relatively small sacrifice from ideal performance.

In summary, the simplest rule of thumb for the table-hopper, irrespective of betting strategy, is to enter in the true count range of +2 to +2.5, depart at around −4, and exit in the range −2 to −2.5. Together these improve the yield ratio by nearly 0.0065, a benefit much greater than from complicated variations in playing the hand to be discussed next in Chap. 5.

Chapter 5
Playing the Hand When the Count and Bet Vary

5.1 Play Strategies That Vary with the Count

As discussed in the previous two chapters, most Players count cards in order to gauge the appropriate amount to bet on the next round. But the best way to *play* that round also depends to some extent on the un-dealt cards, for which counting gives a reasonable indicator; some improvement in performance could be expected from count-guided play adjustments.

Yet any count-dependent play is quite complex. Dozens of play parameters must be specified, such as on which hands to stand or draw, which initial hands to double or split, etc., all depending on Dealer's upcard. Each of these many play parameters may have its own, separate dependence on count, a lengthy set of procedures usually called strategy indices (advocated especially by Schlesinger (2005)).

Because "ideal" play—parameters that change with the individual identities of every card as well as with the count—is generally regarded as too difficult (if not in fact super-human), several classes of simplification are possible with varying levels of precision. For a given count value, the composition-dependent class (as per Sect. 2.2), using the separate identities of just the first two cards plus the upcard, has performance only slightly reduced from ideal; the decision rules for subsequent cards are fixed. The value-dependent class, using just the total value of the initial cards rather than their identities (although still recognizing opportunities to double soft hands and split pairs), is much easier still, but with a small further performance penalty. Basic Strategy, whether Generic or Optimal, is the count-independent instance of this class. This chapter works exclusively with value-dependent strategies, extending them to become count-dependent.

© Springer International Publishing AG, part of Springer Nature 2018
N. R. Werthamer, *Risk and Reward*, https://doi.org/10.1007/978-3-319-91385-8_5

5.1.1 Reconsidering the Counting Vector

A count-dependent extension, however, also calls into question the choice of counting vector(s)! Some choices may be superior for gaining play advantage, even though less desirable for betting purposes, or vice versa. Most ideally, of course, Player would simultaneously maintain all nine separate and distinct counting registers (in effect, knowing the exact composition of the remaining pack); but this is clearly beyond human capability. Less ideal, but difficult nonetheless, is to keep a second, distinct count on which to base play decisions, a possibility suggested by several authors. Even more of a compromise is to use just a single count for both betting and playing decisions, one that optimizes the yield from both together. As an example, advocates of the Hi-Opt counting vector feel it does a better job than the Hi-Lo vector in gaining play advantage. But any of these techniques needs to offer enough yield improvement to justify its added complexity.

The single counting vector that takes optimal advantage of play variation, without any variation in bet, is displayed in Table 5.1. This vector (labeled vp) is much like that in Table 3.2 optimal for bet variation (and labeled vb), except for the striking reversal of sign for aces (card value 1) and the interchanged importance of values 6 and 8. Although vp has non-integer elements and so is not practical for actual counting, several simplifications with integer or half-integer elements suggest themselves, analogous to those of Tables 3.1 and 3.2. The simplest, analogous to the Hi-Opt scheme but with the switch of $j = 3$ and 6 with $j = 7$ and 8, respectively, is also shown in Table 5.1; it has a play correlation of $PC = 0.95$. The Hi-Opt vector itself has $PC = 0.88$, larger than the Hi-Lo vector's $PC = 0.80$ and even the 0.83 of vb; the higher play correlation of the Hi-Opt vector confirms the claim of its proponents that it does a better job than Hi-Lo for play variation.

The reversed effect of aces in playing vs. betting is at the root of the notion in the literature to count aces separately and to use the ace count for improved bet and play variation. This ace adjustment technique, as discussed, for example, by Griffin (1999, pp. 56 & 62), advises Player to start with a counting vector that does not recognize aces, i.e. is "ace-neutral," such as Hi-Opt. At the same time, Player also keeps a separate count of aces—or, more precisely, the difference between the number of aces actually dealt and the number expected based on the depth (a modification necessary to keep the count balanced). Player then subtracts the aces "side-count" from the ace-neutral running count to arrive at the ace-adjusted running count. The latter is then used to guide the bet size, while the ace-neutral count is available (as above) for play variation. The ace-adjusted count has betting

Table 5.1 Counting vector optimal for play variation

	Card value									
Vector element	1	2	3	4	5	6	7	8	9	10
Optimal (vp)	+0.37	+0.33	+0.57	+1.10	+1.52	+0.22	+0.88	+0.66	−0.12	−1.38
simplified	0	0	0	+1	+1	0	+1	+1	0	−1

correlation $BC = 0.95$, an improvement over the 0.87 of Hi-Opt by itself, although not quite up to the 0.97 of Hi-Lo.

Finally, in the more general circumstance where two separate counting vectors are employed, yield improvement is maximized by using vb for bet variation and vp for play variation.

5.1.2 Count Dependence of the Play Parameters

To exhibit play strategy variation in some detail, Table 5.2 shows how the optimal hard stand decision varies with true count, in this example when counting six decks using vb. Optimal play becomes more aggressive (defensive) as the true count is increasingly positive (negative). The other shifts in play strategy, including those for doubles and splits, parallel those shown in Sect. 2.1 for a changing deck number, D; the single-deck strategy there corresponds roughly to that for many decks at a true count of about $+2\frac{1}{2}$. As pointed out by previous authorities, the most significant strategy variations are those that take place near zero true count: the decision whether to stand on 16 or 17 vs. 10, near true count $+0.3$; and whether to stand on 12 or 13 vs. 4, near -0.3. These two shifts are evident in Fig. 8.3. The insurance decision, similarly emphasized by others, is reviewed in the next subsection. The strategy variations for one deck are qualitatively similar to these for six, typically with a shift of about one true count unit.

Nevertheless, play variation by itself improves yield by only a very small amount, much less than that available from optimal betting and/or from table-hopping. At best, the improvement is comparable to that found in Sect. 2.1 for playing Optimal Strategy with a small number of decks rather than Generic. The actual yield improvement depends sensitively on the penetration of the pack: a

Table 5.2 Ranges of true count, between -12 and $+20$, within which it is optimal to stand on a particular hand value or greater, depending on Dealer upcard; six decks

Stand on at least	17	16	15	14	13	12
Upcard						
1	$-12, +9.2$	$+9.4, +10.6$	$+10.8, +13.4$	$+13.6, +19.4$	$+19.6, +20$	
2	$-12, -11.0$	$-10.8, -7.4$	$-7.2, -4.4$	$-4.2, -1.2$	$-1.0, +3.4$	$+3.6, +20$
3	-12	$-11.8, -8.6$	$-8.4, -5.8$	$-5.6, -2.8$	$-2.6, +1.4$	$+1.6, +20$
4		$-12, -10.0$	$-9.8, -7.0$	$-6.8, -4.2$	$-4.0, -0.4$	$-0.2, +20$
5		$-12, -11.6$	$-11.4, -8.8$	$-8.6, -6.0$	$-5.8, -2.4$	$-2.2, +20$
6			$-12, -9.4$	$-9.2, -6.2$	$-6.0, -1.8$	$-1.6, +20$
7	$-12, +9.4$	$+9.6, +11.4$	$+11.6, +15.8$	$+16.0, +20$		
8	$-12, +8.6$	$+8.8, +12.2$	$+12.4, +20$			
9	$-12, +5.4$	$+5.6, +9.8$	$+10.0, +20$			
10	$-12, +0.2$	$+0.4, +4.4$	$+4.6, +11.4$	$+11.6, +20$		

smaller penetration provides less improvement. Furthermore, about 2/3 of this yield improvement is accrued if play is varied based on the counting vector *vb*, optimal for betting, rather than the *vp* optimal for play; this improvement is additive to that resulting from any bet variation.

Considering these results, I feel that the improvement from play variation is too small to be worth the trouble of memorizing much if any of the extensive strategy indices, only a part of which are laid out in Table 5.2. For example, the improvement in yield ratio for the full array is less than +0.001 for a 6-deck pack with *vb* as the counting vector (although it's larger for a single deck). It then follows that a varying play strategy has negligible influence on the evaluation and selection of a counting vector; *vb* alone, or one of its simplifications, is adequate for both tasks. However, even though the issue is dismissed here, its investigation is carried through in Sect. 10.1. The more dedicated Players can decide for themselves whether or not to undertake this complex technique, or even just a subset.

5.1.3 The Insurance Bet

The insurance bet is a play parameter whose count dependence has been of particular interest in the blackjack literature. Whereas insurance is always declined (for any number of decks) with no counting and hence a uniform bet size, authorities assert it becomes favorable for counts exceeding about +4, with either the Hi-Lo or Hi-Opt vectors. The improvement in yield ratio from insurance is said to be as much as +0.006, chiefly because true counts of +4 or more already dictate large bets on a round, prior to the insurance possibility arising. But these assertions are based on a simplification to insurance's expected return which overstates its true worth.

Rather, a more exact analysis shows that the favorable threshold is around +4 only on the reshuffle round, while rising significantly as the depth increases. As a result, farther into the pack the threshold becomes so high that a true count exceeding it becomes highly improbable. The net effect is that even if Player is able to accurately track the correct depth-dependent threshold, the yield is reduced from the amounts usually claimed by roughly 20%–30%, especially with a single deck. And a Player who adheres to the +4 threshold throughout the pack is making some negative expectation bets at larger depths, such that his yield diminishes by as much as another 10%. This added penalty can be largely avoided by shifting the depth-independent threshold upward to +4.5 for multiple decks, +5.5 for single deck. Even so, the ratio of yield to typical bet (the ROI) is barely 0.001 with a single deck, and much less for multiple decks.

These conclusions hold with the use of a counting vector constructed for bet sizing, such as Hi-Lo or Hi-Opt. Griffin (1999, p. 71), and Wong (1994, p. 54), among others, note that a second counting vector, optimized and used exclusively for the insurance decision, might provide additional yield improvement. Based on the same simplification as in the first paragraph, the insurance vector optimal at zero depth is similar to Thorp's Tens count, displayed in Table 5.3. (More precisely,

Table 5.3 Separate counting vector for insurance

Card value	1	2	3	4	5	6	7	8	9	10
Vector element	+2/3	+2/3	+2/3	+2/3	+2/3	+2/3	+2/3	+2/3	+2/3	−3/2

the optimal vector also depends on depth.) The yield ratio resulting from the Tens vector, while dependent on the betting pattern and game parameters, is a roughly 50% improvement over that from the Hi-Lo vector. This seems insufficient to warrant the much greater complexity of tracking a second, separate count.

Overall, insurance does not represent much incremental value, and certainly less than usually claimed: the Hi-Lo or Hi-Opt card counter can either disregard the offer entirely with little loss, or accept the bet for true counts above roughly +5.

5.2 Counter Basic Strategy for the Variable Bettor

Blackjack analysis has been absorbed for decades with the issue of how to simplify or truncate the many strategy changes indicated by a varying card count. An example of a truncated scheme, the Illustrious 18 (or its extension, the Catch 22), appears in Schlesinger (2005). An alternative and much simpler scheme is Counter Basic Strategy (here abbreviated CBS), proposed independently and from different lines of reasoning by Werthamer (2007) and by Marcus (2007).

To begin, focus on the single counting vector optimized for betting. Rather than each play parameter being individually dependent on true count, instead seek the count-*independent* basic strategy that matches the count-dependent one at a "best" choice of true count. The "best" choice (the effective true count) is the one that maximizes Player's yield, recognizing that he is betting different amounts on different hands. Since he makes higher bets for higher true counts, the distribution of his bet sizes is peaked at a true count that is significantly positive. Play that is optimal at or near that peak—not at zero count as prescribed by Basic Strategy—is the "best" choice for the otherwise count-independent CBS.

The methodology and results are illustrated by a few representative examples, adopting *vb*, the counting vector optimal for betting. The same two betting examples are considered as previously, the Weekender and the Lifetimer. Results are displayed, in Table 5.4, only for games with a single deck (where yield is the most sensitive to choice of play strategy) and with four decks (representative of six and eight deck games, as well).

Clearly, CBS increases the yield ratio over that from Optimal Basic by a meaningful amount. The improvement is particularly dramatic for an aggressive bettor like the Lifetimer against a single deck.

It might seem surprising, though, that the yield ratios for CBS and Count-dependent play show no systematic trend. The apparent anomaly can be understood by noting that Count-dependent play maximizes the reshuffle return, at zero depth.

Table 5.4 Yield ratios for Counter Basic, Optimal Basic, and Count-Dependent play strategies

Betting style	Lifetimer		Weekender	
Number of decks	1	4	1	4
Count-independent				
Optimal Basic	+0.076	+0.019	+0.006	+0.008
Counter Basic	+0.091	+0.023	+0.009	+0.011
Count-dependent	+0.083	+0.022	+0.013	+0.011
Effective true count	5.1	3.2	4.6	3.5

Table 5.5 Counter basic strategy, single deck

Upcard	Stand on	Double	Split
A	17;s18	*10*,11	A,8,*9*
2	*12*;s18	9–11;s17,*18*	A,6–9
3	*12*;s18	9–11;*s15,16*,17,18,*19*	A,2,*3*,6–9
4	12;s18	*8*,9–11;s13-18,*19*	A,2,3,6–9
5	12;s18	8–11;s13–18,*19,20*	A,2,3,6–9
6	12;s18	8,9–11;s13–19	A,2,3,6–9
7	17;s18	*9*,10,11	A,2,3,7,8
8	17;s18	10,11	A,8,9
9	*16*;s19	10,11	A,8,9
10	*15*;s19	*10*,11	A,8

Never take insurance.

In contrast, CBS maximizes the yield, the dominant contributions to which come from sizable depths, where the true count is more likely to take on large, positive values and the expected returns and bet sizes are correspondingly larger. In such a regime, the depth dependence of the expected return for a given true count is especially influential in the computational results.

Table 5.5 details the CBS for a single deck, consistent with an effective true count in the vicinity of +5, and identical to the count-dependent but bet-independent play at that true count value (as partly shown in Table 5.2). The parameters that differ from those at zero true count (i.e., Optimal Basic) are in red and italicized.

Chapter 6
Synthesis and Observations

6.1 A Practical, Nearly Optimal Strategy

The preceding chapters have amply illustrated their contention that blackjack is a game that gives Player many options and as a result can be extensively optimized. Reader's interest in the details may well have faded by now! But from the body of material presented a synthesis can be extracted. A simple picture for nearly optimal blackjack emerges, one that greatly enhances Player's yield yet avoids elaborate procedures that don't seem to justify their effort. Let's synthesize that picture, and then compare blackjack as a game and as an enterprise.

The most important message is that *how* one should play blackjack depends on *why* one is playing. The great majority of patrons in casinos are there primarily to have a good time. They enjoy the exciting atmosphere, the pleasure of playing an absorbing game, the thrill of risking a modest amount of money. Although they hope to go home financially ahead, they realize that they probably won't; their priority is to minimize losses and avoid ruin, which would abruptly and prematurely end their session. The remaining patrons, a small minority, are playing primarily to win and are prepared for the hard intellectual effort that winning takes. While the game is just as absorbing and exciting for these Players as for the recreational majority— possibly a good deal more so—it is the financial outcome that motivates them. Their competition against the House is uppermost and they search out every available tool to gain an edge, as well as avoid detection.

6.1.1 Recreational Player

For the recreational Player, I feel that Basic Strategy without counting is adequate. A particular virtue is simplicity: it is easy to learn (or to relearn, after a long hiatus) and to use at the table. Before my occasional casino visits, I practice typically for

© Springer International Publishing AG, part of Springer Nature 2018
N. R. Werthamer, *Risk and Reward*, https://doi.org/10.1007/978-3-319-91385-8_6

about 15 min; a first-timer, with no previous exposure, should be comfortable after about an hour. Making Basic Strategy decisions in real time is not demanding work, even without others at the table to slow down the pace.

But an attribute just as important is that the expected return from Basic Strategy is no worse than roughly -0.006 per hand, even for a shoe with 8 decks and unfavorable rules for splitting and surrender; it is better still with fewer decks (recall Table 2.1) and/or more opportunities for splitting and surrender. Although the House has an edge (except for the one-deck game with DAS and/or resplits, blackjack paying 3 to 2), it is much less unfavorable for Player than any other game in the casino. Certainly anyone tempted to feed the slot machines (return of -0.080 or worse) or to bet on the roulette wheel (-0.053) would be well advised to learn Basic Strategy and head over to a blackjack table!

Generic Strategy, from Table 2.1, uses just the total value of Player's hand and Dealer's upcard. The modifications in play procedure that optimize for the number of decks (Optimal, Table 2.3), or for the specific composition of each hand (Sect. 2.2), improve the expected return so slightly as to not seem worth the recreational Player's effort to memorize. Furthermore, he should bet the same fixed amount on every hand, as he does not have the information needed to rationally vary his bet size.

6.1.2 Competitive Player

For the competitive Player, on the other hand, Generic Strategy and a fixed bet are not enough: he also must track the cards as dealt and adjust his bet size accordingly. With "proper" card counting and betting he can swing his yield from negative to positive: he has an edge over the House, at least in a statistical sense. "Proper" here implies more than just tallying each card as it is dealt: it also involves selecting a counting vector that adequately approximates the optimum, estimating the number of decks remaining in the shoe, and choosing a bet size staircase that appropriately balances Player's reward vs. risk, his greed vs. fear.

The card counting procedure, in review, requires a number of steps. Player must make some choices, a few of which are subjective.

(a) Choose a counting vector. The optimal vector is too difficult for practical use, so an approximation must be adopted. Although several have been discussed, including unbalanced ones, I feel that the Hi-Lo is both simple enough and sufficiently accurate; better still would be either of the Halves vectors, implemented via some card color dependence, for only modest added complexity. The Hi-Lo vector from Sect. 3.2 is reiterated in Table 6.1.

(b) Form the running count. During each shuffle, reset the running count to zero. As each card is dealt, increment the running count by the corresponding element of the counting vector: if a card of value j is dealt, increment the running count by the jth vector element.

Table 6.1 Hi-Lo counting vector

Card value	Ace	2	3	4	5	6	7	8	9	10
Vector element	−1	+1	+1	+1	+1	+1	0	0	0	−1

(c) Convert running count into true count. First, estimate the number of decks remaining un-dealt in the shoe. One way is to track the number of rounds dealt since the last shuffle. If Player is the only one at the table, the number of decks used is about one for each 9 rounds dealt; with two Players it's 6 rounds, with three it's 5, with four it's 4. Then divide the running count by the *unused* decks to arrive at the true count. Since the conversion need only be approximate, estimating the remaining decks just by glancing at the size of the discard pile is quite adequate; rounding off the mental division is also.

(d) Choose a bet staircase (schematically, Fig. 4.1) to convert the true count into the bet size. Specifically, choose the size of the stair-tread, the span of true counts over which the bet size is a given integer multiple of the table minimum; and choose the spread, the ratio of maximum to base bets.

   (d1) The choice, in turn, depends on Player's assessment of his desire for high yield vs. his aversion to ruin. A cautious Player (e.g., the Weekender) might select the MM stair-tread, where his ruin protection is large and his yield is moderately positive. A more aggressive Player, on the other hand (e.g., the Lifetimer), could select the HJY stair-tread, 50% narrower than the MM with a correspondingly steeper staircase, such that his yield is nearly doubled although his ruin protection is rather less. But the Weekender must avoid the temptation to widen his stair-tread much beyond the MM point, i.e., make his staircase even less steep, since the yield then degrades precipitously and soon becomes negative. The effort of card counting is hard to justify if it doesn't give Player an edge!

   (d2) The optimal size of the stair-tread also depends quantitatively on the penetration (as highlighted in Table 4.3). The less of the shoe that is dealt before a reshuffle, the more difficult is Player's achieving a positive yield: the stair-tread must be narrower and the staircase steeper. Yet a bet size that varies substantially from one round to the next can call the House's attention to Player's card counting.

   (d3) In any case, bet the base amount for true counts of +1 or less, because the expected return on those hands is negative.

(e) Play the hand using an appropriate Basic Strategy procedure.

   (e1) For two or more decks, use Generic Strategy as per Table 2.1. Deviating from Generic play based on the true count does improve the yield still further, but so slightly as to question its substantial extra effort. This recommendation is even stronger for the more refined procedures of (i) modifying the counting vector to better accommodate play adjustments or (ii) keeping a second true count exclusively to guide adjustments in play.

Table 6.2 Differences in
1-deck Optimal Strategy from
Generic

Action	Hand value	When Dealer shows
Double	9 and soft 17	2
	Soft 13 and 14	4
	Soft 19	6
	11	Ace
Split	6 + 6	2
	2 + 2	3
	4 + 4	4 with DAS
	6 + 6	7 with DAS
	3 + 3 and 7 + 7	8 with DAS
Stand	Soft 18	Ace
Don't surrender	16	9
	15	10

(e2) Although Generic Strategy is still quite adequate for 1 deck, some Players who do not vary their bets may instead elect to use Optimal Strategy; a few may include some dependence on composition and/or true count. Table 6.2 recaps the modifications to Generic Strategy that convert it into Optimal for a single deck. The expected return is thereby improved by about +0.0003, a very small amount (yet with little or no incremental effort).

(e3) When counting and adjusting bets against a single deck, substitute Counter Basic Strategy, Table 5.4, instead of Generic or Optimal.

The competitive Player can significantly enhance his results in several ways. First of all, he can table-hop, by joining a table only when he's observed its true count to reach the entry threshold, +2 to +3; and/or by leaving before the next reshuffle should the true count subsequently drop to the exit/departure thresholds, −2 to −3. Since the pack after entry is quite favorable and his risk of ruin is otherwise low, he might increase the amount of his base bet and/or, for camouflage, flat bet throughout. Some casinos, however, in an attempt to protect themselves against back-counters, forbid joining a table except during a shuffle. Even in these casinos, though, the exit maneuver (at threshold −1.5 to −2) is quite valuable all by itself.

A second technique is to employ Kelly betting. Player should scale his bet size to his current (rather than initial) capital, by occasionally adjusting his base bet in step with his gains or losses. And if the casino isn't too crowded, he benefits from playing multiple seats, allocating his total bet equally to each. For fastest capital growth the total bet should be a larger proportion of capital than if playing only one seat; but if the proportion is just held the same, the capital variance is reduced even while its growth rate is enhanced.

Team play takes advantage of these techniques to their fullest. A group of optimal Kelly Players who pool and periodically redistribute their capital can come closer to reproducing, and neutralizing, some of the casino's own advantages: multiple (and uncorrelated) tables, large capital reserves, reduced volatility. Table-hopping is also a particularly effective tool for the team that assigns some members as spotters, one to a table; each spotter then signals a "whale" teammate to join and place large bets, when the count at their table has reached the entry threshold. With these procedures, teams can achieve much better results, even per member, than any member could by himself.

The effective yield ratio from counting and optimal betting can be as much as +0.02 or more, even at multi-deck games; and still more when table-hopping. The precise result, though, (as well as the attendant risk of ruin) depends on the number of decks in the shoe and its penetration—set by the House—and especially on the bet stair-tread chosen by Player. Remarkably, Player can always adopt a stair-tread that gives him a positive effective yield, no matter how small the penetration, even though the resulting staircase may be conspicuously steep or, in an extreme case, bump against the table maximum.

Using a strategy with a positive yield does not necessarily mean, however, that Player will win during any given session. The yield is a measure of the *average* outcome, over a very large number of rounds. Over any moderate number of rounds, deviations from that average can be large. For example, consider playing 1000 rounds—over a weekend, say—from a 6-deck shoe and compare the statistical distribution of outcomes for the recreational Player, whose Generic Strategy expected return is about −0.005, with that of the competitive but camouflaged Player, whose yield ratio might be +0.01. *On average* over that session, the recreational Player loses 5 units while the competitive Player wins 10 units.

But the distributions of outcomes in Fig. 6.1, after 1000 rounds, show that each Player is likely to win or lose up to 20 units or more. Even though the peaks (and means) of the distributions are indeed different (−5 vs. +10), the distributions themselves are so broad that the distinction between them is barely perceptible. The

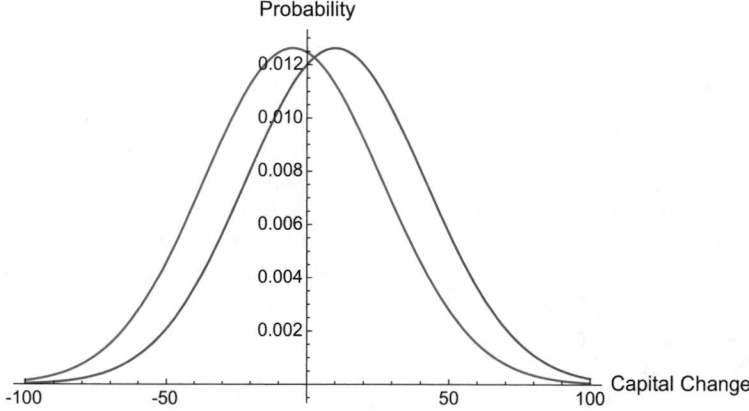

Fig. 6.1 Distribution of outcomes after 1000 rounds. The left-most (red) curve represents the non-counter, the right-most (blue) the counter. To interactively select other values for these parameters, browse to the Wolfram Cloud website www.wolfr.am/BlackjackScience and select the appropriate value. This figure is copyright by N. Richard Werthamer and used by permission

competitive card counter loses in 44% of such sessions; the recreational non-counter wins in 46% of them.

An observer who followed the outcomes of the two Players during that session would be hard pressed to tell who was playing which strategy. Picking as the card counter that Player with the greater outcome would be correct only 57% of the time, not much better than an even chance. Moreover, the observer wouldn't be able to distinguish the two with much certainty until after dozens of such sessions. For example, after 10,000 rounds the distributions are well separated. The observer could then assign strategies to the Players—identify as the card-counter that Player with the better performance—with 82% accuracy. Turning the argument on its side, the Player who counts optimally does not have a substantial assurance of actually winning, or even of outperforming a non-counter, until he has played many thousands of rounds; and even then the assurance is far from total.

Yet the more rounds that are played (whether in one session or in multiple sessions) the greater is the possibility of ruin, terminating the session abruptly and sending Player home unhappy. Since Player's initial capital is finite, it can be consumed by a run of more losing than winning rounds, even if a strategy with positive yield is being used. But ruin does not turn a competitive Player's otherwise winning strategy into a losing one. Even though he may have been ruined in a given session, if he persists and plays many sessions (each, for simplicity of argument, with the same stake and counting strategy), he still has a positive but lesser effective yield as a long-term average. For the recreational Player, ruin lowers his negative yield still further.

The yield reduction factor due to ruin depends on three parameters. The first of these is the coverage (i.e., the ratio of capital to typical bet), which Player can most easily control. The probability of ruin can be decreased by increasing the coverage; a sufficiently high stake relative to typical bet makes ruin negligible. Conversely, a recreational player whose limited budget dictates a small stake should scale down his bets, particularly by choosing a table with the lowest possible minimum.

The second parameter is the ruin protection, associated with the bet staircase selected. Once Player has made his choice between a cautious bet strategy such as the MM point, or a more aggressive one such as the HJY point, the ruin protection is established.

The third parameter is the number of rounds played. Over a short session with at least a moderate coverage, ruin is quite unlikely. As more rounds are played, the risk of ruin grows, although it can be slowed (even if not prevented altogether) by a high enough coverage. In almost all realistic circumstances, ruin will be an occasional occurrence that must be accepted. Player should select a coverage that balances his desire to reduce risk of ruin against a prudent allocation to gambling of a sum he can afford to lose entirely, even if only occasionally.

Money management is the least discussed but nonetheless critical aspect of optimal blackjack (as well as the most complicated to describe!). My sense is that many Players bring too little cash to the casino relative to the size of their bets, and are ruined unnecessarily often. The casino, on the other hand, wins steadily and inexorably because, relative to its patrons, it has vastly more capital, competes at multiple tables, and plays continuously rather than in limited sessions. Thus it can achieve returns much more consistent with its fundamental odds. The facts that those odds favor the House in every game (except, of course, blackjack against a close-to-optimal Player), and that most Players do not play close to optimally in any event, only reinforce the casino's profitability.

6.2   Blackjack as a Recreation vs. a Profession

The picture just described points up the issues facing a Player who seeks not merely to be competitive, but in fact to generate a livelihood exclusively from blackjack. Several well-known books on the subject suggest that blackjack can be a "business" (Revere 1980) or a "profession" (Wong 1994). But let's see what it takes to accomplish this goal. Such a Player is devoting much of his time to working at a blackjack table as well as risking a substantial capital investment. The success of a blackjack professional should be judged on whether his winnings are an adequate return on the combination of his time (his "hourly wage") and on his capital employed (what many companies abbreviate ROCE).

Consider, as an example, a cautious Player with yield ratio of $+0.01$, who plays with a base of \$20 at a typical rate of 100 hands per hour. On average, this Player wins \$20 per hour, a decent if unspectacular wage (it may be more than Dealer's, though!). But such a Player would probably want coverage of around 500, or initial capital of \$10,000. His wage does not leave much room for a high ROCE. A more aggressive Player, on the other hand, with a yield ratio of $+0.02$, might play at a table with a base of \$50. His winnings average \$100 an hour, within the compensation range of other skilled professions. But his stake should probably be \$25,000, and perhaps even higher for added safety. Furthermore, because the investment is at high risk, it should probably be returning at least 20%–40% annually, expensed out of winnings. This requires allocating to ROCE, for instance, at least \$5–\$10 per hour for the first 1000 hours per year.

The real issue, though, for a blackjack professional is the degree of scrutiny he is likely to receive from the House. Although clearly not a topic open to objective analysis, it seems to me that any Player who repeatedly shifts his bet size up and down, regularly plays two dozen hours or so per week at a high stakes table (even though he probably changes casinos frequently), and wins more often than not, is going to be observed closely: that's one reason why casinos have powerful video cameras above every table, are equipped with sophisticated face-recognition software, and share information with other casinos! If the casino then deduces that Player's bet size scales with the true count, his patronage is probably not going to be welcome much longer. In trying to avoid this fate, he will no doubt employ several forms of camouflage. Although camouflage is an art, not a science, whatever techniques he adopts will necessarily degrade his performance. He will feel an ever-present tension between maintaining a decent yield and concealing his counting. Professional blackjack is not a day at the beach!

On the other hand, some experienced casino managers (see, for example, Zender (2008)) point out that many casinos overreact in trying to prevent card counting. For instance, frequent shuffling (low penetration) slows down the game and arguably reduces the casino's overall profitability, even after allowing for some increased winnings from those customers who count cards. Preventing mid-shoe entry is similarly counter-productive. It is possible that expositions of this kind could lead to a relaxation of anti-counter table practices in the future, although some degree of casino scrutiny is likely to remain a given in any event.

The competitive non-professional, however, faces less difficulty. He can lower his bet ramp somewhat to be less conspicuous. He will probably play occasionally, rather than steadily, and perhaps over a shorter session, so he is less likely to be recognized. He may feel comfortable with less camouflage and so his performance may more closely approach the optimum. His average winnings per session will probably be satisfying, although his total winnings over a year should resemble merely an income supplement or a bonus, not his livelihood. Such a Weekender can not only enjoy his session, as much as the non-counting recreational Player, but also earn some compensation for his intellectual skills and effort.

Another contributor to Player's financial performance is casino comps, "gifts" to regulars including a mix of free room, meals, and entertainment (see Scott (2005)). These are likely to comprise a more significant percentage of total compensation for the Weekender, who has less money at stake, than for the Professional, although both are otherwise faced with covering living expenses. On the other hand, a Player who accepts comps is immediately recognized by the casino, which can then evaluate his skill and track his activities.

The non-counting Player should think of blackjack as an entertainment. More likely than not he'll lose some money, but the amount should typically not be greater than what he'd pay for an evening at the theater or concert hall (or, if his taste runs in a different direction, a pro sports game). He'll experience a comparable degree of diversion and, on some occasions, he'll have the added thrill of winning. The Player who counts, on the other hand, is working hard but at a game he enjoys. He should think of blackjack more like his bowling or softball league, with a big cash prize pool. He'll practice to hone his skills and take satisfaction in how well he's using them. Slightly more often than not he'll win some money, and the thrill is more in the looming danger of losing. For each kind of Player, though, there's a strategy optimal for his purpose in playing the game.

Although these realities and choices confront any individual Player, a team is a different story (for particularly vivid ones see Mezrich (2002, 2006) and Anonymous (2015)). A group of card-counters, working collectively, can take advantage of several betting techniques not available to the individual. The team can, in effect, table-hop: they can deploy one or more spotters who flat bet the table minimum, signalling a favorable count to "whales" who enter with big bets and exit should the count turn unfavorable. Team-members' time can be allocated most efficiently with the right ratio of spotters to whales. (The spotters can also depart and switch tables based on the true count threshold, if it wouldn't attract too much attention.) Several whales, always at different tables (or even in different casinos), can occasionally (and out of sight of the casinos, of course) redistribute their capital and Kelly-rescale their bets, gaining almost its full benefit without reduction from any correlation between each other. As long as the team doesn't encounter casino interference, its win rate should be sufficient to cover both a competitive compensation to its members and a reasonable ROCE. A team is a business enterprise, in both principle and fact—with all the attendant management issues of recruiting, training, performance assessment, compensation, assignments of responsibility, investor relations, etc.

A detailed account of one such team has been published by Tilton (2012). He describes a *modus operandi* of working with a single partner, whereby the spotter of the pair uses carefully prearranged signals to convey the current table count to his entering whale. Other aspects of their operation include: confining their gaming almost exclusively to weekends; traveling to locations having several neighboring casinos and alternating among them, to avoid over-exposure; playing at high-limit tables to increase their returns per hour while at the same time being conservative, both in scaling their bet increases and in their capital coverage, to reduce risk of losses; and financing predominantly from a group of non-playing investors who are paid off only after a preset time period spanning multiple playing sessions. Tilton also emphasizes the extensive practice he and his partner devoted to reaching a high level of card counting skill, both in speed and accuracy. (I personally observed him back-counting two tables simultaneously while also chatting with me.) Tilton (private communication) has indicated that their methodology yielded satisfactory hourly wages for the two players as well as attractive returns on capital for their investors.

Part II
Analysis

Chapter 7
Play Strategies

7.1 Basic Strategy, Large Number of Decks

7.1.1 Analytical Framework

Although casinos deal Blackjack from a pack containing up to 8 decks, analysis of the game assuming a much larger number of decks is instructive, regarding both the tools needed and those readily obtainable results that are insensitive to the pack size.

Immediately following a shuffle, the first card drawn from the shoe has a 1 in 13 likelihood of having any specific value between 2 and 9, a 1 in 13 likelihood of being an ace, and a 4 in 13 likelihood of having value 10. In general, the likelihood of a specific value for any subsequently drawn card depends on which cards have previously been drawn; i.e., the composition of the remaining pack fluctuates as cards are dealt from it. However for a shoe with a very large number of decks, the likelihood of drawing a specified value varies only negligibly from that of the first card. Thus, the approximation that those likelihoods are fixed and unvarying is the same as if the shoe contained an infinite number of decks. Griffin (1999) has called the shoe in this limit "non-depleting"; Epstein (2009) says that the draw is "with replacement."

Let $d(i)$ be the likelihood, at any point in the course of dealing the shoe, of drawing the card value i, with $1 \leq i \leq 10$. Also let $d_0(i)$ be the likelihood function at the start of the shoe; clearly $d_0(i) = 1/13$ for $1 \leq i \leq 9$, and $d_0(10) = 4/13$. In the non-depleting shoe approximation, $d(i) = d_0(i)$ for every card drawn.

Next, introduce the pair of integers s and s' to parameterize the hit/stand rules. Let s and s' be the minimum hand values at which Player stands for hard and soft hands, respectively; clearly, if a hand should stand on value s, it should also stand on values $> s$. For Dealer, under typical rules, $s = s' = 17$; in some casinos, $s' = 18$.

Now introduce the conditional probability that a hand valued at c, in the hit range, reaches the value j, in the stand range, after receiving one or more cards. Define $P(j \mid c)$ as the conditional probability for hard hands, with $c < s \leq j \leq 21$, and $P'(j \mid c)$ similarly for soft hands, with $c < s' \leq j \leq 21$. These conditional probabilities satisfy interlinked relations dependent on the size of c. Thus, for c in

© Springer International Publishing AG, part of Springer Nature 2018

N. R. Werthamer, *Risk and Reward*, https://doi.org/10.1007/978-3-319-91385-8_7

the range $11 \leq c \leq s - 1$,

$$P(j \mid c) = d(j - c) + \sum_{k=1}^{s-c-1} d(k) P(j \mid c + k) \qquad (7.1a)$$

The first term corresponds to standing after drawing one card (valuing an ace as 1), the second to drawing the value k (such that $c+k$ is still less than the stand value) and hitting again.

The analogous relation for $P'(j \mid c)$ depends on the size of j relative to s'. A convenient way to accommodate this dependence is to define $\hat{j} \equiv j$ if $s' \leq j \leq 21$, and $\equiv j + 10$ if $s \leq j \leq s' - 1$. Then

$$P'(j \mid c) = d(\hat{j} - c) + \sum_{k=1}^{s'-c-1} d(k) P'(j \mid c + k)$$

$$+ \sum_{k=22-c}^{s-c+9} d(k) P(j \mid c - 10 + k). \qquad (7.1b)$$

The third term corresponds to drawing to the soft hand a card that would bust it except that the ace is revalued from 11 down to 1 so that the hand becomes hard.

Finally, returning to $P(j \mid c)$ in the remaining range of c, namely $2 \leq c \leq 10$,

$$P(j \mid c) = d(j - c) + d(1) P'(j \mid c + 11)$$

$$+ \sum_{k=2}^{s-c-1} d(k) P(j \mid c + k). \qquad (7.1c)$$

The middle term corresponds to drawing an ace, valued at 11 in this case, resulting in a soft hand; and the third term now excludes drawing an ace. Because the form of these relations depends on c and j relative to s and s', both P and P' are also implicitly dependent on those parameters as well.

These relations look at first sight like a large array of simultaneous linear equations for the P and P', a cumbersome task to disentangle. On closer examination, though, they can be seen to be merely recursive: for a given j, s, and s', $P(j \mid c)$ is determined only by the $P(j \mid c')$ for $c' > c$. Also, $P'(j \mid c)$ can be expressed explicitly in terms of the P functions. A computer code to generate the P, given the likelihoods d, is straightforward and executes quickly. Still, the code can be streamlined by first placing the recursive aspect into polynomials $K(m)$, $K^*(m)$ generated from

$$K(m) = \delta(m, 0) + \sum_{n=1}^{m} d(n) K(m - n),$$

$$K^*(m) = \delta(m, 0) + \sum_{n=2}^{m} d(n) K^*(m - n); \qquad (7.2)$$

and then defining auxiliary functions

$$\Pi_\sigma(j \mid c) \equiv \sum_{m=0}^{\sigma-1-c} d(j - c - m) K(m),$$

$$\Pi^*(j \mid c) \equiv \sum_{m=0}^{10-c} d(j - c - m) K^*(m). \tag{7.3}$$

The Kronecker delta functions in Eq. (7.2) are defined as $\delta(i, j) \equiv 1$ for $i = j$, and $\equiv 0$ for $i \neq j$. It is important when interpreting these definitions to recognize that $d(j) = 0$ when j is outside the range $1 \leq j \leq 10$. The expressions (7.1a)–(7.1c) can then be recast, respectively, in the non-recursive form,

$$P(j \mid c) = \Pi_s(j \mid c), 11 \leq c \leq s - 1$$

$$P'(j \mid c) = \Pi_{s'}(\hat{j} \mid c) + \sum_{k=12}^{s-1} \Pi_s(j \mid k) \Pi_{s'}(k + 10 \mid c),$$

$$P(j \mid c) = \Pi^*(j \mid c) + d(1) \sum_{m=0}^{10-c} P'(j \mid c + 11 + m) K^*(m) \tag{7.4}$$

$$+ \sum_{k=11}^{s-1} \Pi_s(j \mid k) \Pi^*(k \mid c), 2 \leq c \leq 10.$$

The conditional probabilities P determine the outcome of Player's contest against Dealer. Whereas for Player the variable c represents the total value of his first two cards (except for split pairs, analyzed later), similar probability functions also apply to Dealer, where the variable instead represents the value of just her upcard; we'll use the notation u instead of c when referencing Dealer. Also, while $P(j \mid c)$ for Player depends implicitly on his stick/hit parameters s, s', the corresponding Dealer's function $P(j \mid u)$ implicitly assumes her mandatory $s = s' = 17$ (or in some casinos $s' = 18$).

7.1.2   Expected Player Return

For a specific c and u, the expected return to Player, labeled $R(c, u)$, is given by the probability that he wins (excluding blackjack), minus the probability that he loses, plus 1.5 times the probability that he (but not Dealer) gets blackjack. Furthermore, Player wins (excluding blackjack) either when his hand total is greater than Dealer's (and neither busts) or when Dealer busts (and Player doesn't). Player loses when his hand total is less than Dealer's (and neither busts), or when he busts (and

Dealer hasn't drawn blackjack), or when Dealer (but not Player) gets blackjack. In the case where Player's initial hand is hard (but not blackjack), this translates algebraically into

$$R(c, u) = \sum_{j=s}^{21} P(j \mid c) \sum_{k=17}^{j-1} P(k \mid u) + \sum_{j=s}^{21} P(j \mid c) \left[1 - \sum_{k=17}^{21} P(k \mid u) \right]$$

$$- \sum_{j=s}^{20} P(j \mid c) \left[\sum_{k=j+1}^{21} P(k \mid u) - P(bj \mid u) \right] \tag{7.5}$$

$$- \left[1 - \sum_{j=s}^{21} P(j \mid c) \right] [1 - P(bj \mid u)] - P(bj \mid u),$$

where bust hands have been eliminated using conservation of probability. Also, $P(bj \mid u)$ is the conditional probability of Dealer getting a blackjack when her upcard has value u; in terms of the conditional probability $P^{(1)}(k \mid c)$ for reaching value k by drawing only one card to value c, $P(bj \mid u) = P^{(1)}(21 \mid u)$. The lengthy expression of Eq. (7.5) for $R(c, u)$ can be manipulated to just

$$R(c, u) = \sum_{j=s}^{21} P(j \mid c) G(j \mid u) - 1, \tag{7.6}$$

where $G(j \mid u)$ is a linear combination of Dealer's conditional probabilities,

$$G(j \mid u) \equiv 2 \left[1 - \sum_{k=j}^{21} P(k \mid u) \right]$$

$$+ P(j \mid u) - \delta(j, 21) P(bj \mid u); \tag{7.7}$$

note for Dealer that $P(k \mid u) = 0$ if $k < 17$.

If Player's initial hand is soft (but not blackjack), Eq. (7.6) for the outcome still applies but with $P(j \mid c)$ replaced by $P'(j \mid c)$; the lower limit on the j sum remains s, not s'. Finally, when Player gets blackjack,

$$R(21, u) = 1.5[1 - P(bj \mid u)]; \tag{7.8}$$

Player wins provided Dealer does not also draw blackjack.

From $R(c, u)$, four more steps are needed to derive Player's net expected return. First, $R(c, u)$ must be modified to account for his options of doubling and splitting. Second, the modified $R(c, u)$ must be weighted over the likelihood of his having a value c on his initial two cards. Third, the hit/stand parameters s, s' must be determined so as to maximize the weighted returns, for each Dealer upcard u; the optimal s, s' are then functions of u. Finally, the maximized return for a given u must be weighted with the likelihood of Dealer having u as her upcard. Let's take each step in turn.

Player *doubles* whenever doubling the bet and drawing only one additional card is more advantageous than his other play options. Thus, replace $R(c, u)$ by $\max\{R(c, u), 2R^{(1)}(c, u)\}$, where $R^{(1)}$ is the return from drawing just one card and is obtained from the expression for R by replacing P by $P^{(1)}$; and $\max\{A, B, ..\}$ denotes the largest of the group A,B,... It may be advantageous to double a hand even if, played normally, it would be in the stand range; for example, Basic Strategy calls for doubling soft 18, rather than standing, when the dealer shows 3 through 6. Another consideration is that doubling becomes an option only *after* Dealer has determined that she does not have blackjack; thus R and $R^{(1)}$ in the doubling expression must be further modified by adding back the term $P(bj \mid u)$, which had been subtracted out in the general expression.

Player *splits pairs* whenever doubling his bet and playing each card of the pair as a separate hand is more advantageous than his other options. The split option occurs when the two-card hand has value $c = 2i$, where i is the value of each individual card. Furthermore, splitting should also be compared with doubling, although splitting almost always yields a superior return to doubling; the exception, for Basic Strategy, is to double two fives rather than split them. Thus, replace $R(2i, u)$ by $\max\{R(2i, u), 2R(i, u), 2R^{(1)}(2i, u)\}$, at least when resplits are not allowed. As in the doubling analysis, the term $P(bj \mid u)$ must be added back into the split hand return, taking account of the knowledge that Dealer does not have blackjack.

Deriving an expression for Player return when resplits are allowed seems, at first sight, to be formidable; in fact, the task is straightforward. Let $R_{sp}(i, u)$ be the expected return for each card of the split pair. The second card drawn will either have value $k \neq i$, with probability $d(k)$, giving a hand with value $i + k$ (when k is not an ace, $i + 11$ when it is) which is played out normally; or the second card will again have value i, with probability $d(i)$, which prompts a resplit with return $2R_{sp}(i, u)$. Thus

$$
\begin{aligned}
R_{sp}(i, u) &= \sum_{k \neq i} d(k) R(i + k, u) + 2d(i) R_{sp}(i, u) \\
&= \left(\sum_{k} d(k) R(i + k, u) - d(i) R(2i, u) \right) / (1 - 2d(i)).
\end{aligned}
\tag{7.9}
$$

But $\sum_{k} d(k) R(i+k, u) = R(i, u)$. Thus the modification for resplits is that $2R(i, u)$ is replaced by $2(R(i, u) - d(i) R(2i, u))/(1 - 2d(i))$. This modified expression exceeds $R(2i, u)$ (i.e., splitting is favored) precisely when $2R(i, u)$ itself does; hence, the option of splitting a pair is decided independently of whether or not resplits are allowed. Recall, of course, that aces cannot be resplit.

Once $R(c, u)$ has been modified for doubles and splits, it must be weighted with the likelihood of Player's hand having value c. The weighted return, denoted

$R(u)$, is

$$R(u) = \sum_{c_1=2}^{11} d(c_1) \sum_{c_2=2}^{11} d(c_2) R(c_1 + c_2, u). \tag{7.10}$$

This sum must then be broken up into groups, each group being specific for whether or not the two cards form a pair without aces, or contain one or two aces without blackjack, or form blackjack. Comparing the $R(u)$ for all possible Player hit/stand strategies yields the optimized return, $\max_{s,s'}\{R(u)\}$, for the specified Dealer upcard. The optimized return is then weighted over the upcard likelihoods, leading to an expected Player return

$$R = \sum_{u=1}^{10} d(u) \max_{s,s'}\{R(u)\}. \tag{7.11}$$

Combining all four steps into a single expression, lengthy but more explicit,

$$R = \sum_{u=1}^{10} d(u) \max_{s,s'} \left\{ \begin{array}{l} 2 \sum_{c_1=3}^{10} \sum_{c_2=2}^{c_1-1} d(c_1)d(c_2) \max\left\{ R(c_1+c_2, u), 2R^{(1)}(c_1+c_2, u) \right\} \\[2mm] +2 \sum_{c_1=2}^{9} d(c_1)d(1) \max_{s,s'}\left\{ R'(c_1+11, u), 2R'^{(1)}(c_1+11, u) \right\} \\[2mm] + \sum_{i=2}^{10} (d(i))^2 \max\left\{ R(2i, u), 2R_{sp}(i, u), 2R^{(1)}(2i, u) \right\} \\[2mm] +(d(1))^2 \max\left\{ R(12, u), 2R^{(1)}(11, u) \right\} \\[2mm] -(1 - 2d(1)d(10)) P(bj \mid u) + 2d(1)d(10) R(21, u) \end{array} \right\} \tag{7.12}$$

The terms containing $d(c_1)d(c_2)$ and $d(c_1)d(1)$ correspond to hard and soft hands, respectively, evaluated for doubling; the terms with $(d(i))^2$ and $(d(1))^2$ correspond to pairs of non-aces and of aces, respectively, evaluated for splitting and doubling; and the term with $d(1)d(10)$ corresponds to blackjack.

7.1.3 Frequency of Tied Hands

Although most hands either win or lose, some hands tie and no money is exchanged. This occurs on nearly 10% of all rounds. Although this fact may not seem significant, since ties do not affect the player's financial situation, the frequency of ties does enter the picture in Sect. 4.1, which considers optimal betting and money management. Ties also affect the variance of the return, considered in the next subsection.

In a formalism suitable for the variance as well, begin with a compressed notation. Firstly, let k index the final value of Dealer's hand, itemized as: stand on 17 to 21 but not blackjack, bust on 22 and higher, or blackjack on two-card 21. Further, let $\xi_k(\omega)$ be the probability (also depending implicitly on u) that Player wins/loses ω units against Dealer outcome k; ω takes on the values (assuming no DAS or resplits) $0, \pm 1, +3/2, \pm 2$. Then the probability of tying is

$$T = \sum_u d(u) \sum_k P(k \mid u)\xi_k(0). \tag{7.13}$$

Now re-expand the notation to make explicit Player's initial hand value c, with probability $P(c) = \sum_{c_1,c_2} d(c_1)d(c_2)\delta(c_1 + c_2, c)$. Distinguish those values that are doubled as c_d; those that are split as $c_s = 2i$; and those that are blackjack. Denote the probability of Player winning/losing ω units contingent on c and k as $\xi_k(\omega \mid c)$. Then

$$T = \sum_u d(u) \sum_k P(k \mid u) \left[\begin{array}{l} \sum_c P(c)\xi_k(0 \mid c) + \sum_{c_d} P(c_d)\xi_k(0 \mid c_d) \\[2mm] +P(bj)\xi_k(0 \mid bj) \\[2mm] + \sum_{c_s} P(c_s)\left(\xi_k(0 \mid i)^2 + 2\xi_k(+1 \mid i)\xi_k(-1 \mid i)\right) \end{array} \right]. \tag{7.14}$$

Note (third line) that a split pair can tie not just by each of the two resulting hands tying separately but also by one winning and the other losing.

Finally, substitute for the $\xi_k(\omega \mid c)$ in terms of the conditional probabilities and use conservation of probability to eliminate bust terms, yielding

$$T = \sum_u d(u) \left[\sum_c P(c) \sum_\kappa P(\kappa \mid u)P(\kappa \mid c) + \sum_{c_d} P(c_d) \sum_\kappa P(\kappa \mid u)P^{(1)}(\kappa \mid c_d) \right.$$

$$+P(bj \mid u)P(bj) + \sum_{c_s} P(c_s) \left[P(\kappa \mid i)(2P(\kappa \mid i) - 1)P(\kappa \mid u) \right.$$

$$\left. \left. + \sum_{j=s}^{21} P(j \mid i)\left((1 - P(j \mid i))G(j \mid u) - 2\sum_{j'=s}^{21} P(j' \mid i)G(j' \mid u)\right) \right] \right] \tag{7.15}$$

where κ runs over those stand values common to both Dealer and Player, $\max\{17, s\} \leq \kappa \leq 21$. Computation of this expression for Basic Strategy yields $T = 0.0982$. For the rounds that aren't tied, Player wins about 42.7% of the total and loses about 47.5%; the fact that some winning hands pay more than 1 to 1, particularly blackjack at 3 to 2, substantially reduces the House's edge over Player from the 4.8% indicated by the difference of these two numbers.

It proves convenient for the next subsection to distinguish notationally the separate contributions to T from doubled, split and blackjack hands:

$$T \equiv \sum_{c} P(c)t + \sum_{c_d} P(c_d)t_d + \sum_{c_s} P(c_s)t_s + P(bj)t_{bj}. \qquad (7.16)$$

7.1.4 Multiple Simultaneous Hands: Return, Variance, and Covariance

Intuition suggests that the expected return from H hands played simultaneously, each with a unit bet, should be identical to the return from a single hand played with a bet of H. But a mathematical first look is less obvious. Although H unit-bet hands played *sequentially* obviously have the same total return as a single hand with a bet of H units, each of these hands competes against a separate Dealer hand. Multiple hands played *simultaneously* all compete against the *same* Dealer hand, and so might acquire some correlation that would affect the results.

Nevertheless, intuition is correct for the expected return. Carrying over the notation of the previous subsection, the expected return and the mean square return from a single hand are

$$\langle R_1 \rangle = \sum_u d(u) \sum_k P(k \mid u) \sum_\omega \omega \xi_k(\omega),$$
$$\langle R_1^2 \rangle = \sum_u d(u) \sum_k P(k \mid u) \sum_\omega \omega^2 \xi_k(\omega), \qquad (7.17)$$

so that the variance of the return is $\sigma_1^2 = \langle R_1^2 \rangle - \langle R_1 \rangle^2$. The total expected return from H hands against a single Dealer hand is

$$\langle R_H \rangle = \sum_u d(u) \sum_k P(k \mid u) \left(\prod_{h=1}^{H} \left(\sum_{\omega_h} \xi_k(\omega_h) \right) \right) \left(\sum_{h=1}^{H} \omega_h \right), \qquad (7.18)$$

which, since $\sum_\omega \xi_k(\omega) = 1$ for every k, easily reduces to $\langle R_H \rangle = H \langle R_1 \rangle$ as claimed.

The distinction between simultaneous and sequential play of multiple hands sharpens when turning to the second moment (or mean square deviation, or variance) of the return. Putting pair splitting aside for the moment, the analog to Eq. (7.18) is

$$\langle R_H^2 \rangle = \sum_u d(u) \sum_k P(k \mid u) \left(\prod_{h=1}^{H} \left(\sum_{\omega_H} \xi_k(\omega_h) \right) \right) \left(\sum_{h=1}^{H} \omega_h \right)^2, \qquad (7.19)$$

which can be manipulated into the form

$$
\begin{aligned}
\sigma_H^2 &\equiv \left\langle R_H^2 \right\rangle - \langle R_H \rangle^2 \\
&= H \left(\left\langle R_1^2 \right\rangle - \langle R_1 \rangle^2 \right) \\
&\quad + H(H-1) \left[\sum_u d(u) \sum_k P(k \mid u) \left(\sum_\omega \omega \, \xi_k(\omega) \right)^2 - \langle R_1 \rangle^2 \right] \\
&\equiv H \sigma_1^2 + H(H-1)\Gamma.
\end{aligned}
\tag{7.20}
$$

The covariance Γ doesn't vanish because all Player hands are played against the same Dealer hand, and so are correlated. As a result, the variance of the H Player hands is greater than H times that of a single hand.

When pair splitting is re-introduced, the expressions grow longer but the key results,

$$
\langle R_H \rangle = H \langle R_1 \rangle,
$$

$$
\left\langle R_H^2 \right\rangle - \langle R_H \rangle^2 = H \sigma_1^2 + H(H-1)\Gamma,
\tag{7.21}
$$

remain the same. The single hand variance and the covariance generalize to

$$
\begin{aligned}
\sigma_1^2 &= \sum_u d(u) \sum_k P(k \mid u) \left[\sum_\omega \omega^2 \xi_k(\omega) \right. \\
&\quad \left. + \sum_{\omega_1,\omega_2} (\omega_1 + \omega_2)^2 \xi_k(\omega_1)\xi_k(\omega_2) \right] - \langle R_1 \rangle^2, \\
\Gamma &= \sum_u d(u) \sum_k P(k \mid u) \left[\sum_\omega \omega \xi_k(\omega) \right. \\
&\quad \left. + \sum_{\omega_1,\omega_2} (\omega_1 + \omega_2)\xi_k(\omega_1)\xi_k(\omega_2) \right]^2 - \langle R_1 \rangle^2,
\end{aligned}
\tag{7.22}
$$

where here the ω sums are over non-split results and the ω_1, ω_2 sums are over the results of the two split hands.

Substitution for the $\xi_k(\omega)$ and manipulation as before (making use of the notation of Eq. (7.16), as predicted) leads to the compact expression

$$
\begin{aligned}
\sigma_1^2 + \langle R_1 \rangle^2 &= \sum_c P(c)(1-t) + 4 \sum_{cd} P(c_d)(1-t_d) \\
&\quad + (9/4) P(bj)(1-t_{bj}) + 4 \sum_{c_s} P(c_s)[1-t_s \\
&\quad - (3/2) \sum_u d(u) \sum_\kappa P(\kappa \mid u) P(\kappa \mid i)(1 - P(\kappa \mid i))].
\end{aligned}
\tag{7.23}
$$

Thus the probability of tying enters essentially into the variance, which can be characterized as the sum of the squares of the amounts won or lost, each weighted by their probability excluding ties. The computation of the variance is thus an easy extension of that for the probability of ties, resulting in $\sigma_1^2 = 1.261$, or $\sigma_1 = 1.123$.

The covariance expression in Eq. (7.22) can also be manipulated into a computation-friendly form, but most compactly with a somewhat different notation from that of the variance. First define the auxiliary quantities

$$\wp(j) \equiv \sum_c P(c)P(j \mid c)$$

$$+ 2 \sum_{c_d} P(c_d)P^{(1)}(j \mid c_d) + 2 \sum_{c_s} P(c_s)P(j \mid i);$$

$$\wp \equiv \sum_j \wp(j) = \sum_c P(c) + 2 \sum_{c_d} P(c_d) + 2 \sum_{c_s} P(c_s);$$

$$\wp^*(j) \equiv \sum_i \varepsilon_{i,j}\wp(i), \quad \varepsilon_{i,j} \equiv +1, i > j; -1, i < j; 0, i = j.$$

(7.24)

Then in this notation the expected return of Eq. (7.17) can be recast as

$$\langle R_1 \rangle = \sum_u d(u) \left[\sum_j \wp(j)G(j \mid u) - \wp \sum_k P(k \mid u) \right] + R_{bj}. \qquad (7.25)$$

The covariance analogously becomes

$$\Gamma = \sum_u d(u) \left[\sum_j \left(\begin{matrix} 2\wp(j)\wp^*(j)G(j \mid u) \\ -\wp(j)^2 P(j \mid u) \end{matrix} \right) + \wp^2 \sum_k P(k \mid u) \right]$$
$$+ [1.5P(bj) - \langle R_1 \rangle]^2 + P(bj)[1.5P(bj)]^2. \qquad (7.26)$$

Computation of this form gives $\Gamma = 0.472$. Previous authorities have either estimated the covariance as $\Gamma \approx 0.50$ (Epstein (2009), p. 285; Griffin (1999), p. 142) or evaluated it using simulations as 0.47 (Wong (1994), Table 85 (no DAS), p. 203).

7.1.5　Expected Number of Cards Used per Round

The conditional probabilities P can also be used to derive the expected number of cards used in playing a round. While this aspect is not directly relevant to Player's financial return, as noted before, it is useful in Chap. 3, which discusses card counting methods.

Looking back for a moment at the Eqs. (7.1a)–(7.1c) satisfied by $P(j \mid c)$ and $P'(j \mid c)$, note that they could be iterated to express $P(j \mid c)$ as a power series in d. But for each such term, the number of cards drawn is just the number of factors of d. Thus expressions for $N(j \mid c)$ and $N'(j \mid c)$, the expected number of cards used in moving the hand value from c to $j \leq 21$, can be obtained by replacing in $P(j \mid c)$ (at least symbolically) every factor of d^n by nd^n. Such a replacement can be accomplished concretely by functionally differentiating the equations for $P(j \mid c)$, $P'(j \mid c)$ with respect to d, since

$$d \frac{\delta}{\delta d} d^n = n d^n. \tag{7.27}$$

Functionally differentiating Eq. (7.1a), for example, yields the recursion relation

$$N(j \mid c) = d \frac{\delta}{\delta d} P(j \mid c)$$
$$= P(j \mid c) + \sum_{k=1}^{s-c-1} d(k) N(j \mid c + k), \tag{7.28}$$

which is similar to Eq. (7.1a) except for the change in the inhomogeneous term. Coding these relations to compute N is only a minor supplement to the code for computing P itself.

However, a consideration not present in analyzing the return is that cards are also used for hands that bust and must be included alongside those used for hands that stand. But Eq. (7.1a), and by extension Eq. (7.25), are also valid for $21 < j < s+10$. Furthermore, the quantity of interest is $N(c)$, the number of cards drawn to a given initial hand, irrespective of the final hand value or whether or not the hand busts. Then summing Eq. (7.28) over all $s \leq j \leq s+9$ gives the very simple result,

$$N(c) = 1 + \sum_{k=1}^{s-c-1} d(k) N(c + k) = \sum_{k=1}^{s-c-1} K(m). \tag{7.29}$$

In the remaining ranges of c, $N(c)$ satisfies the corresponding analogs of Eq. (7.4). In particular, by introducing analogs to Π, Π' of Eq. (7.3),

$$\Omega_\sigma(c) \equiv \sum_{m=0}^{\sigma-1-c} K(m), \quad \Omega^*(c) \equiv \sum_{m=0}^{10-c} K^*(m), \tag{7.30}$$

N and N' can be expressed as

$$N(c) = \Omega_s(c),$$

$$N'(c) = \Omega_{s'}(c) + \sum_{k=12}^{s-1} \Omega_s(k)\Pi_{s'}(k + 10 \mid c),$$

$$N(c) = \Omega^*(c) + d(1)\sum_{m=0}^{10-c} N'(c + 11 + m)K^*(m) \tag{7.31}$$

$$+ \sum_{k=11}^{s-1} \Omega_s(k)\Pi^*(k \mid c).$$

The next steps in deriving the total number of cards used by Player, N_p, are to modify $N(c)$ and $N'(c)$ for doubles and splits, which depend on Dealer's upcard u; to weight the result by the probabilities of c and u; and then to add 2 for the initial two cards in Player's hand. The result is

$$N_p = 2 + \sum_{u=1}^{10} d(u)[\sum_{c_1=3}^{10} \sum_{c_2=2}^{c_1-1} d(c_1)d(c_2)[N(c_1 + c_2), 1]$$

$$+ \sum_{c_1=2}^{9} d(c_1)d(11)[N'(c_1 + 11), 1] \tag{7.32}$$

$$+ \sum_{i=2}^{10} d(i)^2[N(2i), 2 + 2N_{sp}(i)]$$

$$+ d(1)^2[N'(12), 2]].$$

The number of cards used in splitting is given by $N_{sp}(i) = N(i)$, without resplits, and $= (N(i) - d(i) N(2i))/(1 - 2 d(i))$ with resplits.

A similar analysis holds for Dealer, except that she neither doubles nor splits; her corresponding quantity $N(u)$ is weighted over the probability of drawing the upcard u; and her initial card number is just 1, not 2. Hence

$$N_d = 1 + \sum_{u=2}^{11} d(u) N(u). \tag{7.33}$$

Computation of these expressions gives $N_p = 2.75$, $N_d = 2.91$.

But the analysis so far is based on considering Dealer's and Player's actions as uncorrelated. In casino practice, though, there are several interrelationships between the two, particularly when only a single Player is at the table: if either Dealer or Player has blackjack, the other does not draw additional cards; and, if Player busts, Dealer similarly does not draw additional cards. Incorporating these realities, the respective card numbers change to $N_p = 2.72$ and $N_d = 2.76$, with a 5.48 total average number of cards used per round. For multiple Players at the table, the average number of cards used by Dealer and by each Player rises slightly.

7.2   Basic Strategy, Small Number of Decks

7.2.1   Analytical Framework; Reordering Theorem

Now turn to treating the game with a small number of decks. An exact analysis accounting for depletion is daunting: rather than the compact recursion relations of the previous section, now the likelihood of drawing a card at any point in the play depends on precisely which cards were already drawn, both in that round and in all previous rounds of the shoe. As a consequence, every hand allowed by the rules is to be catalogued and its probability evaluated; for hands that split pairs, the possibilities reach into the billions. Here we concentrate on just the first round after a shuffle; subsequent rounds, further into the pack, are treated, in the next Chapter.

The number of distinct hands needing individual consideration can be reduced considerably by recognizing that:

1. hands for which Player and/or Dealer bust can be eliminated (using conservation of probability) in favor of a more complex weighting of those that don't, generalizing the result of Eqs. (7.6)–(7.7);
2. rounds that tie do not contribute to the expected return;
3. all hands that have the same cards—even though in differing sequence—with each conforming to the rules, need evaluation only once with their probability weighted by their multiplicity.

The last of these claims is not obvious. Its justification is sufficiently significant to warrant calling it a theorem. Begin the proof by denoting the number of decks used as D, so that immediately after the shuffle the pack contains $52D$ cards. Assume that M cards have been dealt in playing one or more previous rounds; despite claiming that only the first round after a shuffle is being considered here, the extension to $M > 0$ provides a base for another important result proved in Sect. 8.1.2. Then $52D - M$ cards remain in the pack at the start of dealing the current round. Its first card (dealt to Player) has a value c_1 with likelihood $d^{(1)}(c_1) = d(c_1)$; although $d(c_1)$ immediately after the shuffle is unchanged from its value $d_0(c_1)$, this is generalized in Sect. 8.1. The second card (dealt to Dealer) has value u with likelihood

$$d^{(2)}(u; c_1) = \frac{(52D - M)d(u) - \delta(u, c_1)}{52D - M - 1}$$

$$= d(u) + \frac{d(u) - \delta(u, c_1)}{52D - M - 1}.$$

(7.34)

The numerator of the first expression gives the number of cards remaining in the pack with value u, reduced by one if u equals c_1; while the denominator is the total number of cards remaining after the first is dealt. The third card (dealt to Player again) has value c_2 with likelihood

$$d^{(3)}(c_2; c_1, u) = d(c_2) + \frac{2d(c_2) - \delta(c_2, u) - \delta(c_2, c_1)}{52D - M - 2}. \qquad (7.35)$$

Extending this sequence to the mth card with value j_m gives the likelihood expression

$$d^{(m)}(j_m; j_1, \ldots, j_{m-1}) = \frac{d(j_m) - \varepsilon \sum_{k=1}^{m-1} \delta(j_m, j_k)}{1 - \varepsilon(m-1)}, \qquad (7.36)$$

where $\varepsilon \equiv 1/(52D - M)$, the inverse number of cards remaining, can be considered small and a possible expansion parameter.

The probability of any given hand is just the product of the likelihoods of each card in the hand, i.e., a product of factors $d^{(m)}$, noting that each factor depends on the identities of the cards preceding it. But the order of the cards can be interchanged without affecting the hand's probability! Specifically, from Eq. (7.36) it can be shown that two adjacent factors satisfy

$$d^{(m)}(j_m; j_1, \cdots, j_{m-1}) d^{(m+1)}(j_{m+1}; j_1, \cdots, j_m)$$
$$= d^{(m)}(j_{m+1}; j_1, \cdots, j_{m-1}) \, d^{(m+1)}(j_m; j_1, \cdots, j_{m-1}, j_{m+1})' \qquad (7.37)$$

which demonstrates that interchanging the pair of cards j_m, j_{m+1} does not alter the probability of the hand they are in. By iteration, any reordering of a hand's cards leaves its probability unchanged. This result is so important for the subsequent analysis that we elevate its status, calling it the Reordering Theorem.

Reducing the number of distinct hands using the Reordering Theorem is complicated, however, by the need to exclude hands not conforming to the rules. An example shows the non-trivial nature of the symmetry involved: the card sequence 6 followed by ace followed by 7 (here denoted 6+A+7) is a hand with value 14, a legitimate Player stand for Dealer upcards 2 through 6. Player always draws a third card to his soft 6+A. However, the rearranged sequence 6+7+A, with the identical value and probability, is not a legitimate hand because Player stands after the 6+7 and doesn't draw the third card.

More generally, an algorithm to generate the irreducible set of distinct, legitimate hands follows the form of Eqs. (7.1a)–(7.1c). Recognize that if each of these equations were fully expanded, removing their recursiveness, the result would be a lengthy sum of terms, each of which is the product of several factors of d. Although each product gives a hand probability applicable only for an infinite deck, the structure of the product is useful for a finite deck: the arguments of the d factors correctly indicate the values of the cards making up that hand. In this way, Eqs. (7.1) serve as a framework from which to generate all applicable hands, both for Dealer and Player. The roster of cards making up a hand can then be processed in a more complex way to compute its probability for any finite number of decks. For example, consider a pack of 6 decks totaling 312 cards, from which a hand of 5 cards is dealt consisting of three 2s and two 5s, making a hand value of 16; if the hand directly follows a reshuffle, its probability is $(24!/21!)(24!/22!)/(312!/307!)$ and not simply $(1/13)^5$.

Another element in the analysis that warrants an extended description stems from the procedure of Dealer peeking to determine whether she holds blackjack. As discussed in Sect. 1.2, in most casinos when Dealer's upcard is 10 or ace she immediately checks her downcard (or peeks) for blackjack, in which event there is no further play of the hand. Only if the downcard is, respectively, not ace or 10 does play proceed. Thus the analysis is predicated, when $u = 10$ or 11, on the downcard having value $u_2 \neq u^*$, where $u^* \equiv 21-u$. Thus with the probability of the downcard u_2 prior to peeking being $d^{(4)}(u_2; k_1, u, k_2)$, the conditional probability after peeking becomes $d^{(4)}(u_2; k_1, u, k_2)/C^{(4)}(k_1, u, k_2)$, where

$$C^{(4)}(k_1, u, k_2) \equiv 1 - (\delta(u, 1) + \delta(u, 10))d^{(4)}(u^*; k_1, u, k_2). \qquad (7.38)$$

This result mirrors that of Epstein (1995), p. 224 (this material has been omitted from his 2nd edition). Note that the $d^{(4)}$ factor in $C^{(4)}$ anchors the downcard, after peeking, as the fourth card dealt in the round.

An efficient computational program for finite numbers of decks takes further advantage of the Reordering Theorem to modify the order of play: shifting the entire Dealer hand to being dealt first, before any actions by the Player. Thus all possible Dealer hands are generated and stored first. Then as each Player hand is generated, its return is computed by folding its probability against the function Eq. (7.7) of the array of Dealer hands.

This program is both exact and feasible, but needs some further elaboration when pair splitting is considered. In that case, since the precise probability of the second split hand depends on the cards dealt to the first, the number of distinct combined hands becomes roughly the square of the number without splitting; the computational effort to proceed along this line becomes too vast. Unlike those earlier authors who evaluate split pairs by resorting to simulation, we here take Player's return from splitting to be just twice that of playing the first of the split hands alone. As an example, when the pair 8+8 is split, instead of playing out each 8 in turn, we take the return to be twice that of playing out only one of the 8s.

This stratagem, although it may seem to be an approximation, is actually exact: a consequence of again applying the Reordering Theorem. Rigorously, the expected return from the split pair is the sum of the returns from each of the split hands individually, played in sequence and followed by Dealer's hand. But rather than calculating the return from the first hand as though it were dealt first, followed by the second, followed by Dealer's, the Theorem shows the equivalence with dealing to the first hand, then to Dealer's, and then to the second hand. The return from the second hand, similarly, is equivalent with dealing to it first, followed by Dealer's hand and then to the first hand. The sum resulting from the reordering is just the stratagem described. The equivalence further extends to multiple resplits, as well as to cases with a small numbers of decks where one or more card values may become exhausted before the end of the round.

7.2.2 Expected Return and Optimal Basic Strategy, vs. Number of Decks

The outcome of the computations with Optimal Basic Strategy for the expected return from the first round following a shuffle is listed in Table 2.2 of Part I. The four rule possibilities for split pairs are included: with and without DAS, with and without resplits. The returns are also plotted in Fig. 2.1. In each case the points are closely linear in the inverse number of decks, with a barely perceptible upward quadratic; this functionality is justified from a more general perspective in Sect. 8.1. The linear dependence is fitted in the form $R_D \approx R_\infty + R'/52D$, where the slope parameters R' are also shown in Table 2.2.

Further, shown here in Table 7.1, are results for the variation with number of decks of the probability of tying and of the return's variance and covariance. These were given in the previous section in the non-depleting pack approximation, but they can also be computed for any finite number of decks. If Generic Strategy is assumed throughout, then T changes only slightly with D while σ_1^2 and Γ change even less.

Computation of the covariance—the correlation of returns from two hands played simultaneously against the same Dealer hand—is feasible only with an approximation: that the probability of the second hand reaching any given final count is the same as that of the first hand. Since the covariance depends only slightly on pack size anyway, this approximation likely has little effect on the results.

Optimal Basic Strategy for various numbers of decks is listed in Table 2.3. There are more changes in strategy for smaller D, which contribute slightly to the dependence of the expected return displayed in Table 2.2. But the optimization has been carried out with the Basic Strategy constraints:

Table 7.1 Ties, variance, and covariance vs. number of decks

D	T	σ^2	Γ
∞	0.0982	1.2611	0.4719
8	0.0975	1.2608	0.4715
6	0.0972	1.2607	0.4713
4	0.0968	1.2605	0.4710
2	0.0954	1.2599	0.4700
1	0.0926	1.2587	0.4681
Linear fit	$0.0982 - 0.292/52D$	$1.2611 - 0.125/52D$	$0.4719 - 0.201/52D$

- that the parameters for standing are a function only of Dealer's upcard, not the individual identities of Player's first two cards, other than to distinguish pairs from non-pairs and hard from soft;
- that the doubling specifications are a function only of Dealer's upcard *and* the combined value of Player's initial two cards, distinguishing hard from soft;
- and that each split hand is played with the same strategy as an unsplit one.

7.2.3 Surrender; Insurance

Surrender is easily analyzed: any two-card hand whose expected return is less than -0.5 should be surrendered, when that option is available. For a pack with more than 2 decks, this is the case for hard hands with value 16 (other than the 8+8 pair, which should still be split) against upcards 9, 10 and ace; and also value 15 against upcard 10. When surrender is available, the overall expected return with a large number of decks improves by 0.0009, a meaningful amount, although as shown in Table 2.2 the improvement declines substantially with decreasing D. Note that surrender is indicated mostly when Dealer's upcard is 10 or ace and she hasn't drawn blackjack; proper account of this added information is necessary in computing the expected return from Player's two-card hand.

With respect to Insurance, Player has the option of side-betting an additional half his primary bet when Dealer shows an ace. The side bet pays off at 2 to 1 if Dealer's hole card is a 10, independently of all other aspects of the hand. Hence Player's expected return from Insurance on the reshuffle round is just

$$R_I = d_0(11)\,(2d_0(10) - (1 - d_0(10)))\big/2, \tag{7.39}$$

which evaluates to approximately -0.003. Since the return is negative, and the bet is not contingent on any other aspect of play, the Insurance offer should be declined. However, when a card count is kept circumstances do arise, addressed in Sect. 10.1.4, when Insurance develops a positive return and the option might be accepted.

7.3 Play Parameters Dependent on Identities of Initial Cards

In the previous section the stand parameters were optimized as a function only of Dealer's upcard. Thus Player stands on any hard hand value $\geq s$ and draws to any value $< s$, independent of the cards making up that hand, and correspondingly s' for soft hands. But he might also use knowledge of the identities of his first two cards. For example, an initial hand value of 13 might be comprised of either 6+7 or 5+8. With upcards such that hitting the 6+7 is optimal, would Player improve his return by instead standing on 5+8? As another example, can the return be increased by

doubling 3+5 but not 2+6? Once these questions are answered—yes in both cases—
it is then natural to ask by how much.

Thus now allow s,s' to depend on the specific identities of the initial two cards,
not just Dealer's upcard; also, separately optimize the play of each two-card hand
formed from the first of a split pair and the next card drawn to it. (A still further
generalization, termed "ideal" in Chap. 10, would be to let s and s' depend on *each
and every* card in the hand, not just the first two—in other words, to re-optimize
the decision of hitting vs. standing each and every time another card is drawn—but
the resulting strategy rules would be overwhelmingly complex and with very little
improvement in expected return to show for it.)

The results of this composition-dependent strategy analysis are presented in
Table 7.2, along with the consequent improvement in expected return (roughly
independent of the rules for playing split hands) over Optimal Basic Strategy.

The stand profile vs. upcard 10 is truly remarkable! When no account is taken of
Player's initial hand value (i.e., in Optimal Basic Strategy) the stand parameter is s

Table 7.2 Composition-dependent play strategy

D	Dealer upcard	Exceptions to Optimal Basic Strategy, Table 2.3	Increase in $R \times 1000$
8	10	Stand on 16[a]	0.017
	Ace	Stand on soft 18 for 2+2, 3+2, 3+3	
6	4	Hit 2+10	0.028
	10	Stand on 16[a]; don't surrender 8+7	
	Ace	Stand on soft 18 for 2+2, 3+2, 3+3	
4	2	Double split 2+7, 6+ace, 7+2, 7+ace	0.052
	4	Hit 2+10	
	10	Stand on 16[a]; don't surrender 8+7	
	Ace	Stand on soft 18 for soft $c<17$, hard $c<8$	
2	2	Double split 6+ace, 7+ace	0.132
	3	Stand on 12 for split 6, 7, 8	
	4	Hit 2+10; double split 3+ace	
	6	Double split 3+5	
	10	Stand on 16[a]; surrender 6+9, 5+10	
	Ace	Stand on soft 19 for soft $c>16$, hard $c>17$; don't double 3+8, 2+9	
1	2	Hit 3+10	0.370
	3	Stand on 12 but hit 3+9, 2+10	
	4	Hit 2+10	
	5	Don't double 2+6	
	6	Hit 2+10; don't double 2+6	
	10	Stand on 14 for 7+7, surrender if allowed; Stand on 15 for 5+4, 5+5; stand on 16[a]; surrender 6+9, 5+10	
	Ace	Stand on 19 with 6+ace; don't surrender 7+9	

[a]But stand on 17 with 9+7 and 10+6 if surrender isn't available; and also with 8+2 and all other
hands with a 6, 10+3 for $D \geq 2$, 7+3 for $D \geq 4$, 10+2 with $D \geq 6$, and 3+3 with $D \geq 8$

= 17 for all D, even $D = 1$. Nevertheless, $s = 16$ is in fact optimal for the majority of initial hands, even for as many as 8 decks! This seeming contradiction is resolved by discovering that $s = 16$, when optimal, is favored only very slightly over $s = 17$; and that $s = 17$, when favored over 16, does so by a margin one to two orders of magnitude larger. Thus $s = 17$ emerges from a weighted average over all hands. Nearly 70% of the improvement is reproduced by a simplified rule: stand on every hard hand with value 16 or more except draw to 6+10 (and split 8+8, of course). Also remarkable against upcard 10, when $D = 1$, is the rule to stand on 7+7 (or preferably surrender, if possible).

Similarly unexpected is the soft stand parameter $s' = 18$ against Dealer's ace for hands 2+2, 3+2 and 3+3, for all $D \le 8$. Although Optimal Basic Strategy calls for stand on soft 18 against ace for 2 or fewer decks, the persistence of this effect for small two-card hand values up to as many as 8 decks is surprising. Nevertheless, the result is quite plausible: those few hands that begin with 2+2, 2+3 or 3+3 vs. ace, and conclude with soft 18, have so depleted the pack in low-valued cards (ace, 2 or 3) that the chance of drawing another such, to reach 19, 20, or 21, is much reduced.

The improvement in expected return resulting from the composition-dependent strategy is insignificant except for $D = 1$, validating the remark in Sect. 2.2 that the complexity of a composition-dependent strategy might only be justified for a single-deck game. At the other extreme, for a large number of decks the computation demonstrates the absence of any composition dependence: in that case Basic Strategy is also optimal even over the much broader class of strategies based on the individual identities of the first three cards.

As noted earlier, the value of Surrender, and the hands that should be surrendered, shrink from that of Sect. 2.1 as the number of decks decreases and Player's return correspondingly improves; the surrender increment for $D = 1$ is a mere 0.00024. Composition-dependent changes to Sect. 2.1 surrender guidelines are that 7+8 should not be surrendered against upcard 10 for $D < 7$; nor 16-valued hands vs. 9 for $D < 3$; nor 7+9 vs. ace for $D = 1$.

7.3.1 Comparison with Previous Authorities

A number of previously published analyses give exact strategies for the first round after a shuffle with specific numbers of decks. I'm aware of those by Manson et al. (1975), Griffin (1999), Epstein (1995), Revere (1980), Thorp (1966) and Wong (1994); more recently an exhaustive tabulation by Cacarulo has been reported in Appendix A of Schlesinger (2005). A close comparison with these results is worthwhile.

The work of Manson et al. (1975) purports to treat exactly the reshuffle round from a four-deck shoe, although in fact they cope with pair splitting by resorting to computer simulation. Wong (1994), furthermore, obtains strategy results entirely by simulation. While such a method can be accurate to within narrow statistical limits, it offers few insights into questions other than the specific ones that have been programmed in. Most of the other studies employ a computational approach apparently originated by Julian Braun. Although no study describes its methodology in detail, Griffin (1999), p.154 and p. 172 and Epstein (1995), pp. 223–224 give the fullest outline. A key idea is to compare, for each possible Player hand, the expected return from standing with that from drawing an additional card, and to select the option whose return is larger. Each hand is linked to those with higher values by a recursion similar to that expressed by Eqs. (7.1)–(7.4).

The approach here, on the other hand, is to compute the expected return weighted over all hands with a given initial value, and to maximize the return with respect to the stand parameters s, s'. (Recall that these are the values at or above which hard or soft hands, respectively, should stand.) But these two criteria can be shown to be identical in outcome: for example, the hard hand value for which standing gives a larger return than hitting, while hands with value lower by one are better off hitting than standing, is just the value s for which the expected return is a local maximum.

Some overlaps among the parameters in these publications permit direct inter-comparisons. In particular, Manson et al. (1975) and Griffin (1999), pp. 173 ff, give composition-dependent strategy results for four decks that agree completely with each other. Griffin (1999) and Wong (1994) also agree on strategy with a single deck. However, as noted earlier, Manson et al. (1975) investigate pair splitting using a simulation technique; Griffin (1999), pp.154–157 implies that this part of their work is inaccurate and provides corrected results. Various discrepancies among the other earlier studies can be found.

Relative to them, it is widely agreed that Cacarulo's more recent results in Schlesinger (2005) are the most reliable. He gives strategies and expected returns for 1, 2, 4, 6 and 8 decks, with both "total-dependent (TD)" and "composition-dependent (CD)" hands. However, he uses slightly different definitions of these than we do: his "TD" includes certain multi-card decisions that we consider only as CD, such as hit-or-stand on 16 vs. Dealer's 10; and his "CD" excludes some multi-card decisions, such as double-or-hit a two-card split hand distinct from if it weren't split. Our strategy results agree exactly with Cacarulo's for two-card hit-or-stand decisions, even though he defines away from "CD" some of the multi-card rules shown in Table 7.2 and includes others into "TD." As a further consequence of these definition differences, the expected returns exhibited in Tables 2.2 and 2.4, although within two to three significant figures of his in Table A1 of Schlesinger (2005), p. 394, are slightly smaller than for "TD" and slightly larger than for "CD."

Chapter 8
Card Counting

8.1 Analytical Framework

The analysis of Optimal Basic Strategy in Sect. 7.2 considered only the round immediately following a shuffle, so that the first card drawn had value j with likelihood $d_0(j)$. But for subsequent rounds, deeper into the pack, the distribution of the remaining cards varies and so do the likelihoods $d(j)$.

Begin by assuming a pack of D decks, with $v_j \equiv 52Dd_0(j)$ cards of value j, $1 \leq j \leq 10$. Further assume that m_j cards of value j are drawn at random from the pack, M cards altogether. At this point the distribution of cards $\tilde{m}_j \equiv v_j - m_j$ remaining in the pack, with total $\tilde{M} \equiv 52D - M$, is the multinomial (the "multivariate hypergeometric," mathematicians might say),

$$p(\tilde{\boldsymbol{m}}) = \delta \left(\sum_j \tilde{m}_j, \tilde{M} \right) \frac{\tilde{M}! \left(52D - \tilde{M} \right)!}{(52D)!} \prod_j \frac{v_j!}{\tilde{m}_j! \left(v_j - \tilde{m}_j \right)!}. \tag{8.1}$$

The distribution is normalized, in that $\prod_{j=1}^{10} \sum_{\tilde{m}_j=0}^{v_j} p(\tilde{\boldsymbol{m}}) = 1$. But it is usually easier to work with the distribution's asymptotic limit for large v_j.

8.1.1 Asymptotic Distribution

The likelihood of drawing value j on the next card is $d(j) = \tilde{m}_j / \tilde{M}$. Use the notation $\mathbf{d}$ for the 10-dimensional vector of likelihoods, and $\rho(\mathbf{d})$ for its distribution. Also denote by $f \equiv M/52D$ the fraction (or depth) of the shoe already dealt, and define the quantity $\Delta \equiv (f/52D(1-f))^{1/2}$; because of the large factor $52\,D$ in its denominator, Δ can conveniently be regarded as a small parameter. Further assume the pack has a sufficiently large number of cards of each value (i.e., $4\,D >> 1$) that $\mathbf{d}$ can be closely approximated as a continuous variable, permitting differential

© Springer International Publishing AG, part of Springer Nature 2018
N. R. Werthamer, *Risk and Reward*, https://doi.org/10.1007/978-3-319-91385-8_8

and integral operations on functions of it. In fact, the results emerging from this continuous variable approach are good approximations even for small numbers of decks. As derived in the Appendix 1, $\rho(\mathbf{d})$ is then given asymptotically by

$$\rho(\mathbf{d}) = \sqrt{2\pi}\,\Delta\,\delta\left(\sum_{j=1}^{10} d(j) - 1\right)$$

$$\times \prod_{j=1}^{10} \frac{1}{\sqrt{2\pi\,d_0(j)}\,\Delta} \exp\left[-\frac{1}{2}\frac{(d(j) - d_0(j))^2}{d_0(j)\,\Delta^2}\right] \tag{8.2}$$

Thus each of the card values is Gaussian (i.e., bell curve) distributed about its mean, and Δ scales the width of the distribution. The width increases with increasing f and with decreasing D. The distribution is normalized, in the sense that $\int d\mathbf{d}\,\rho(\mathbf{d}) = 1$, with a mean $\langle\mathbf{d}\rangle \equiv \int d\mathbf{d}\,\mathbf{d}\,\rho(\mathbf{d}) = \mathbf{d}_0$.

The discussion of Chap. 3 involved just a single counting vector $\boldsymbol{\alpha}$. We gain insight by generalizing to multiple counting vectors, designated $\boldsymbol{\alpha}_\lambda$, where $1 \leq \lambda \leq \Lambda \leq 9$. This generalization extends to the extreme of counting each card value separately, corresponding to $\Lambda = 9$ (the 10th is fixed by $\sum_j d(j) = 1$), which is the maximum amount of information that can be gained from scanning the dealt cards. Multiple counting vectors leads to multiple true counts γ_λ:

$$\gamma_\lambda = \sum_{i=1}^{10} \frac{\alpha_\lambda(i)\,m\,(i)}{D - M/52} = \frac{\boldsymbol{\alpha}_\lambda \cdot \mathbf{m}}{D\,(1 - f)}. \tag{8.3}$$

Based on knowledge of the true counts, the distribution of the likelihoods is narrowed; since $\mathbf{d} = (52D\mathbf{d}_0 - \mathbf{m})\big/(52D - M)$, the distribution of $\mathbf{d}$ conditional on γ is

$$\rho(\mathbf{d}|\gamma) = \frac{\rho(\mathbf{d})}{p(\gamma)}\prod_{\lambda=1}^{\Lambda}\delta\left(\gamma_\lambda + \boldsymbol{\alpha}_\lambda \cdot \left[52(\mathbf{d} - \mathbf{d}_0) - \frac{M\,\mathbf{d}_0}{D\,(1 - f)}\right]\right), \tag{8.4}$$

where δ is the Dirac delta function such that, for any function $f(x)$, $\int f(x)\,\delta(x)dx = f(0)$; and $p(\gamma)$ is the normalization factor ensuring that $\int d\mathbf{d}\,\rho(\mathbf{d}|\gamma) = 1$ for every γ. Furthermore, $p(\gamma)$ is the probability distribution of γ, such that $\int d\gamma\,p(\gamma)\,\rho(\mathbf{d}|\gamma) = \rho(\mathbf{d})$.

Moments of the conditional distribution can be extracted by substituting Eq. (8.2) into Eq. (8.4) and taking the integral over $\mathbf{d}$; then replacing each delta function by its integral representation, $\delta(x) = (2\pi)^{-1}\int_{-\infty}^{\infty} dy\,\exp(i\,x\,y)$ and performing the various Gaussian integrals, beginning with the $\mathbf{d}$ variables, using as a prototype $\int_{-\infty}^{\infty} dy\,\exp(ixy - y^2/2) = \sqrt{2\pi}\exp(-x^2/2)$. The manipulations are straightforward except when the matrix of inner products

$$A^2_{\lambda,\lambda'} \equiv (52\,\Delta)^2 \left[\sum_j d_0(j)\,\alpha_\lambda(j)\,\alpha_{\lambda'}(j) - \sum_{j,k} d_0(j)\,d_0(k)\,\alpha_\lambda(j)\,\alpha_{\lambda'}(k) \right]$$

(8.5)

is encountered. Although in many situations the counting vectors might be orthogonal, so that $\mathbf{A}^2$ is a diagonal matrix, it is informative to consider the more general case of non-orthogonal vectors. Then the zeroth moment of the conditional distribution leads to an expression for $p(\gamma)$,

$$p(\gamma) = \frac{1}{\mathrm{Det}\mathbf{A}} \left(\frac{1}{\sqrt{2\pi}} \right)^\Lambda \exp\left\{ -\frac{1}{2} \sum_{\lambda,\lambda'} X_\lambda\, A^{-2}_{\lambda,\lambda'}\, X_{\lambda'} \right\},$$

(8.6)

where $X_\lambda \equiv \gamma_\lambda - M\boldsymbol{\alpha}_\lambda \cdot \mathbf{d}_0 / D\,(1-f)$ and Det is the determinant operation. Expression (8.6) shows that γ is also Gaussian distributed, not surprisingly since γ and $\mathbf{d}$ are linearly related. Also, the mean of γ_λ is just $\langle \gamma_\lambda \rangle = M\boldsymbol{\alpha}_\lambda \cdot \mathbf{d}_0 / D\,(1-f)$, and its second moments are

$$\langle \gamma_\lambda \gamma_{\lambda'} \rangle - \langle \gamma_\lambda \rangle \langle \gamma_{\lambda'} \rangle \equiv \int d\gamma\, p(\gamma)\,(\gamma_\lambda - \langle \gamma_\lambda \rangle)(\gamma_{\lambda'} - \langle \gamma_{\lambda'} \rangle)$$

$$= A^2_{\lambda,\lambda'}.$$

(8.7)

The objective for this machinery is to derive the expected return as a function of γ, and to optimize it with respect to the counting vectors. To begin, it proves convenient to make a change of variables to what we'll see is a normalized, scale-invariant basis. Designating the new variables with a caret, define

$$\hat{d}(i) \equiv (d_0(i))^{1/2},$$

$$\hat{\boldsymbol{\alpha}}_\lambda(i) \equiv 52\,\hat{d}(i) \sum_{\lambda'} A^{-1}_{\lambda,\lambda'}\,(\alpha_{\lambda'}(i) - \boldsymbol{\alpha}_{\lambda'} \cdot \mathbf{d}_0),$$

$$\hat{\gamma}_\lambda \equiv \Delta \sum_{\lambda'} A^{-1}_{\lambda,\lambda'}\,(\gamma_{\lambda'} - \langle \gamma_{\lambda'} \rangle).$$

(8.8)

In this "caret" representation, the $\Lambda+1$ variables are indeed orthonormal:

$$\left| \hat{d} \right|^2 = 1, \quad \hat{d} \cdot \hat{\boldsymbol{\alpha}}_\lambda = 0, \quad \hat{\boldsymbol{\alpha}}_\lambda \cdot \hat{\boldsymbol{\alpha}}_{\lambda'} = \delta(\lambda, \lambda').$$

(8.9)

Also, the $\hat{\gamma}_\lambda$ are invariant against a scale change in the counting vectors $\boldsymbol{\alpha}_\lambda$. Furthermore, it is convenient to introduce both the derivative operator $\hat{\nabla}$, where

$$\hat{\nabla}_i \equiv \hat{d}(i) \left(\frac{\partial}{\partial d_0(i)} - \sum_j d_0(j) \frac{\partial}{\partial d_0(j)} \right)$$

(8.10)

is such that $\hat{\boldsymbol{d}} \cdot \hat{\nabla} = 0$, as well as the tensor $\hat{\boldsymbol{D}}$, where

$$\hat{D}_{i,j} \equiv \delta\,(i,\,j) - \hat{d}(i)\,\hat{d}(j) - \sum_{\lambda=1}^{\Lambda} \hat{\boldsymbol{\alpha}}_\lambda(i)\,\hat{\boldsymbol{\alpha}}_\lambda(j). \tag{8.11}$$

$\hat{\boldsymbol{D}}$ is a projection matrix: it has $\Lambda+1$ vanishing eigenvalues, corresponding to its eigenvectors $\hat{\boldsymbol{d}}$, $\hat{\boldsymbol{\alpha}}_\lambda$, with the remaining eigenvalues equal to 1. When $\Lambda = 9$, these eigenvectors are also complete, satisfying Eq. (8.11) with $\hat{\boldsymbol{D}} = 0$. Eigenmodes and related concepts in matrix algebra are sketched in Appendix 2; see also, e.g., Shores (2007).

Returning to Eq. (8.4) for the conditional distribution, the caret notation allows exhibiting its first two moments in an especially simple form:

$$\langle d(i)\rangle_\gamma \equiv \int d\mathbf{d}\, d(i)\, \rho(\mathbf{d}|\gamma)$$

$$= d_0(i) - \sum_{\lambda=1}^{\Lambda} \hat{\gamma}_\lambda\, \hat{d}(i)\, \hat{\boldsymbol{\alpha}}_\lambda(i), \tag{8.12}$$

and

$$\Big\langle \big(d(i) - \langle d(i)\rangle_\gamma\big)\big(d(j) - \langle d(j)\rangle_\gamma\big)\Big\rangle_\gamma = \Delta^2\,\hat{d}(i)\,\hat{d}(j)\,\hat{D}_{i,j}. \tag{8.13}$$

Since the conditional distribution is Gaussian, all of its higher moments can be expressed in terms of just these two.

8.1.2   Expected Return; Invariance Theorem

Turn now to Player's return when counting cards. Section 7.1 derived the expected return $R(\mathbf{d})$ from a reshuffle round with likelihoods $\mathbf{d}_0$ at its start. Here, knowing the true counts γ, the likelihood $\mathbf{d}$ for the first card departs from $\mathbf{d}_0$ as per Eq. (8.12).

The expected return on a hand, at a depth f and conditional on true counts γ, is given by

$$\langle R_f(\mathbf{d})\rangle_\gamma = \big\langle \exp\big[(\mathbf{d} - \langle\mathbf{d}\rangle_\gamma) \cdot \nabla\big]\big\rangle_\gamma\, R_f\big(\langle\mathbf{d}\rangle_\gamma\big)$$

$$= \exp\Big((1/2)\,\Delta^2\,\hat{\boldsymbol{D}} : \hat{\nabla}\hat{\nabla}\Big)\, R_f\big(\langle\mathbf{d}\rangle_\gamma\big)$$

$$= \exp\left[\frac{1}{2}\Delta^2\Big(\hat{\nabla}^2 - \sum_\lambda (\hat{\boldsymbol{\alpha}}_\lambda \cdot \hat{\nabla})^2\Big) - \sum_\lambda \hat{\gamma}_\lambda\,\hat{\boldsymbol{\alpha}}_\lambda \cdot \hat{\nabla}\right] R_f\,(\mathbf{d}_0)\,,$$

$$\tag{8.14}$$

where the second line uses Eq. (8.13) and the third uses Eqs. (8.11) and (8.12). The exponential of derivatives in Eq. (8.14) is symbolic. If $R_f(\mathbf{d}_0)$ had a Fourier transform, the exponential operator on it would become a Gaussian integration over the transform. Here, where $R_f(\mathbf{d}_0)$ is only known numerically, the operator is defined in practice through its series expansion. Note that only the linear combinations of card values that are *not* counted enter the diffusion-like exponential factor, as per Eq. (8.11), whereas those that *are* counted enter only in $\langle \mathbf{d} \rangle_\gamma$, as per Eq. (8.12).

Expression 8.14 can be substantially simplified, in the absence of counting, via the important result $\langle R_f(\mathbf{d}) \rangle = R_0(\mathbf{d}_0)$. This implies that the effect of fluctuations in the composition of a depleting pack exactly cancels the change in its depth from zero, as alluded to in the previous Chapter. A significant further consequence is that without counting the expected return from a round is totally independent of its depth, a result termed the Invariance Theorem by Thorp (2000).

The proof begins by observing that the expected return is the return from each specific configuration of cards dealt to Player and Dealer, weighted by the probability of that configuration being dealt, and summed over all allowable configurations. Thus, like the approach of Sect. 7.2,

$$R_f(\mathbf{d}) = \sum_j \Xi(\mathbf{j}) \prod_{m=1}^N d^{(m)}(j_m; j_1, \ldots, j_{m-1}),\qquad(8.15)$$

where $\Xi(\mathbf{j})$ is the return from the configuration $\mathbf{j} = j_1, \cdots, j_N$ of an N card hand, and the product $d^{(1)} \cdots d^{(N)}$ is the probability of that configuration.

Next, rearrange the cards of the hand, after M cards have previously been dealt, into a canonical form where the n_1 aces are dealt first, the n_2 twos next, etc., so that $\sum_k n_k = N$. The hand can be characterized just by the numbers $\mathbf{n}$ rather than by the full set $\mathbf{j}$. Then by the Reordering Theorem of Sect. 7.2.1,

$$P(\mathbf{n}; M) \equiv \left\langle \prod_{m=1}^N d^{(m)}(j_m; j_1, \ldots, j_{m-1}) \right\rangle$$

$$= \left\langle \prod_{k=1}^{10} \prod_{\mu_k=0}^{n_k-1} \frac{v_k - m_k - \mu_k}{\tilde{M} - \sum_{l=1}^{k-1} n_l - \mu_k} \right\rangle.\qquad(8.16)$$

To evaluate the expectation in Eq. (8.16), begin by folding its argument against the exact multinomial distribution Eq. (8.1) rather than the asymptotic limit Eq. (8.2). Then replace the Kronecker delta in Eq. (8.1) by its Fourier representation, leading to

$$P(\mathbf{n}; M\ ) = \frac{M!\tilde{M}!}{(52D)!} \int_0^{2\pi} \frac{d\varphi}{2\pi} \exp(-i\, M\, \varphi)$$

$$\times \prod_{k=1}^{10} \left[\sum_{m_k=0}^{v_k} \left(\prod_{\mu_k=0}^{n_k-1} \frac{v_k - m_k - \mu_k}{\tilde{M} - \sum_{l=1}^{k-1} n_l - \mu_k} \right) \frac{v_k!\exp(i\, m_k\, \varphi)}{m_k!(v_k - m_k)!} \right]$$

$$= \frac{M!(\tilde{M} - N)!}{(52D)!} \int_0^{2\pi} \frac{d\varphi}{2\pi} \exp(-i\, M\, \varphi)$$

$$\times \prod_{k=1}^{10} \left[\prod_{\mu_k=0}^{n_k-1} \left(v_k - \frac{1}{i} \frac{\partial}{\partial\varphi} - \mu_k \right) \right] \left(1 + e^{i\,\varphi} \right)^{v_k}.$$

$$(8.17)$$

The multiple derivative operations can be carried out by iteration of the prototype $(a + i\partial/\partial\varphi)\left(1 + e^{i\,\varphi}\right)^a = a\left(1 + e^{i\,\varphi}\right)^{a-1}$, such that each iteration reduces the exponent a by one. Then straightforward manipulations give

$$P(\mathbf{n}; M) = \frac{M!(\tilde{M} - N)!}{(52D)!} \int_0^{2\pi} \frac{d\varphi}{2\pi} \exp(-i\, M\, \varphi)$$

$$\times \prod_{j=1}^{10} \left[\frac{v_j!}{(v_j - n_j)!} \left(1 + e^{i\,\varphi} \right)^{v_j - n_j} \right] \qquad (8.18)$$

$$= \frac{(52D - N)!}{(52D)!} \prod_{j=1}^{10} \frac{v_j!}{(v_j - n_j)!}$$

$$= P(\mathbf{n}; 0), \quad \text{QED}.$$

Thus the probability of drawing any hand is independent of the depth at which it is dealt, and so $\langle R_f(\mathbf{d}) \rangle = R_0(\mathbf{d}_0)$ as claimed! Furthermore, since the proof is valid for any function $\Xi\,(\mathbf{j})$, it is true not just for blackjack but for a wide variety of other card games, baccarat for example. Thorp (2000) obtained this important result from an entirely different line of reasoning.

8.1.3 Hermite Series

Substituting this result back into Eq. (8.14) gives the simpler expression,

$$\left\langle R_f(\mathbf{d}) \right\rangle_\gamma = \exp\left[-\sum_\lambda \left(\hat{\gamma}_\lambda \, \hat{\alpha}_\lambda \cdot \hat{\nabla} + \frac{1}{2} \Delta^2 (\hat{\alpha}_\lambda \cdot \hat{\nabla})^2 \right) \right] R_0(\mathbf{d}_0). \qquad (8.19)$$

Furthermore, using Eq. (8.7) in the caret form $\left\langle \hat{\gamma}_\lambda \, \hat{\gamma}_{\lambda'} \right\rangle = \Delta^2 \delta(\lambda, \lambda')$, the exponential in Eq. (8.19) can be expressed as a power series in the small parameter Δ^2; in the case of a single counting vector, the coefficients are the Hermite polynomials H_n:

$$\left\langle R_f(\mathbf{d}) \right\rangle_\gamma = \sum_{n=0}^{\infty} \frac{\Delta^n}{n!} H_n \left(-\sum_\lambda \frac{\hat{\gamma}_\lambda}{\Delta} \hat{\alpha}_\lambda \cdot \hat{\nabla} \right) R_0(\mathbf{d}_0). \qquad (8.20)$$

These polynomials are mutually orthogonal against the weighting factor $p(\gamma)$, in that for an arbitrary Λ-dimensional vector X,

$$\left\langle H_n \left(\frac{\hat{\gamma} \cdot X}{\Delta} \right) H_m \left(\frac{\hat{\gamma} \cdot X}{\Delta} \right) \right\rangle = \delta(n, m)\, n! \ |X|^{2n} \qquad (8.21)$$

Also, using multiple integrations by parts on any differentiable function $J(\hat{\gamma})$, they can be shown to satisfy

$$\left\langle J(\hat{\gamma}) \, H_n \left(\frac{\hat{\gamma} \cdot X}{\Delta} \right) \right\rangle = (X \cdot)^n \left\langle \frac{\partial^n J(\hat{\gamma})}{\partial \hat{\gamma}^n} \right\rangle, \qquad (8.22)$$

as well as the derivative relation

$$\partial H_n \left(\frac{\hat{\gamma} \cdot X}{\Delta} \right) \Big/ \partial X = n \left(\frac{\hat{\gamma}}{\Delta} \right) H_{n-1} \left(\frac{\hat{\gamma} \cdot X}{\Delta} \right). \qquad (8.23)$$

Some low-order instances in the scalar case are $H_0(z) = 1$, $H_1(z) = z$, $H_2(z) = z^2 - 1$, $H_3(z) = z^3 - 3z$.

8.2 Expected Return at Nonzero Depth

The Hermite series gives a precise way to extend results away from the zero-depth reshuffle round to finite depths. It is particularly valuable in simplifying computations of yield and related functions, to be pursued in subsequent chapters. Note that the expected return when counting is also dependent on depth, as per Eq. (8.19), even though without counting the Invariance Theorem proves its depth independence. Further note that $R_0(\mathbf{d}_0)$, in Eqs. (8.19) and (8.20), also depends implicitly on the Basic Strategy play parameters.

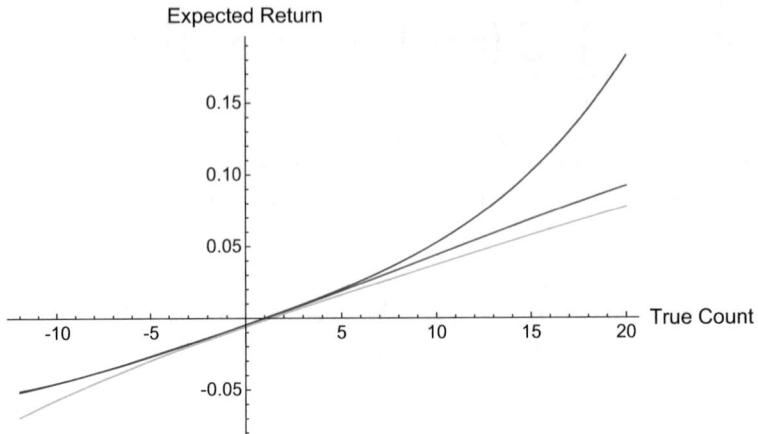

Fig. 8.1 Expected return vs. true count, reshuffle round, six decks: upper two curves for Count-Dependent play, the topmost (blue) with DAS, the middle (red) without; the lowest curve (green) is for count-independent Optimal Basic play

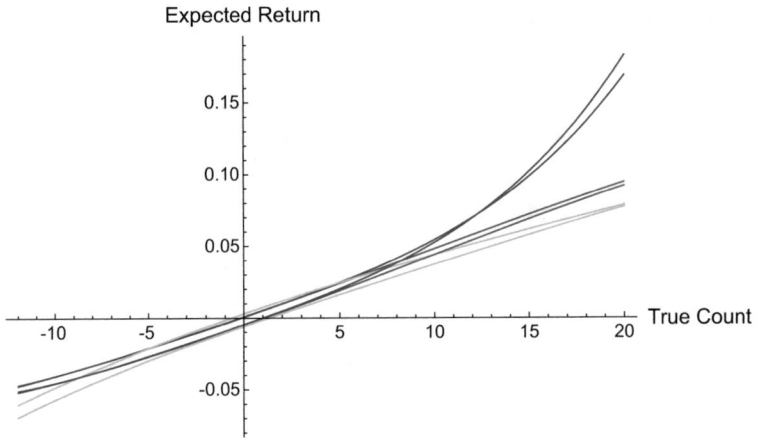

Fig. 8.2 Expected return vs. true count as in Fig. 8.1, but superimposing onto the six-deck results those for single deck as well, the latter with a shape similar to the former but a slight upward shift

In the case of a single counting vector, the reshuffle return $R_0\left(\langle\mathbf{d}\rangle_\gamma\right)$ can be computed as a function of $\hat{\gamma}$, starting with Eq. (7.12) for R_0 and Eq. (8.12) for $\langle\mathbf{d}\rangle_\gamma$. Results for expected return over the true count range of -12 to $+20$ are shown in Fig. 8.1, for both Optimal Basic play and Count-Dependent play, six decks, with and without DAS. The count-independent Optimal Basic curve is convex downward, whereas the adjustments of Count-Dependent play as per Sect. 5.1.2 improve the expected return to concave upward. Although the curves of Fig. 8.1 are with resplits allowed, the shape is similar without resplits. Figure 8.2 illustrates the similarity of the results for single deck by superimposing them onto those for six decks.

Table 8.1 Parameters in polynomial fits to expected return vs. true count, Optimal Basic (OB) and Count-Dependent (C-D) play strategies

	OB					C-D				
	$10^3 c_0$	$10^3 c_1$	$10^5 c_2$	$10^6 c_3$	$10^8 c_4$	$10^3 c_0$	$10^3 c_1$	$10^5 c_2$	$10^6 c_3$	$10^8 c_4$
$D = 1$	-0.53	4.57	-4.04	0.92	-1.68	-1.08	4.65	4.21	-2.36	1.58
$D = 2$	-3.98	4.63	-3.90	0.90	-1.70	-4.47	4.68	4.80	-2.56	1.89
$D = 4$	-5.66	4.58	-3.70	1.52	-3.81	-5.74	4.62	4.31	-2.01	2.15
$D = 6$	-6.22	4.59	-3.66	1.54	-3.95	-6.29	4.62	4.45	-2.06	2.19
$D = 8$	-6.50	4.60	-3.64	1.56	-4.03	-6.57	4.62	4.51	-2.09	2.21

These graphs are much like those of Fig. 3.1, but with true count expanded out beyond the linear regime. Note that the improvement in expected return from Count-Dependent play results predominantly from large true counts that occur only with low probability.

These zero-depth curves can be fitted adequately by a low-order polynomial in true count, $R_0\left(\langle \mathbf{d} \rangle_\gamma\right) \cong \sum_n c_n \hat{\gamma}^n$, based on $1/52D$ being a small parameter. A cubic works very well for Optimal Basic play while, in the case of Count-Dependent play with DAS, accuracy can be improved by adding a quartic term. In fact, Figs. 8.1 and 8.2 show *both* the data *and* its polynomial fits, a superposition that is undetectable because the differences are much less than the width of the curves! The resulting coefficients c_n are given in Table 8.1. Then a comparable fit to the return at a general depth is just $\langle R_f(\mathbf{d}) \rangle_\gamma \cong \sum_n c_n \Delta^n H_n\left(\hat{\gamma}/\Delta\right)$. This dramatic simplification is exploited repeatedly in Chaps. 9 and 10.

Comparison of the series expansion just above to that of Eq. (8.20) is facilitated in the small Δ^2 limit, since the large argument limit of $H_n(z)$ is just z^n. The comparison reveals that c_n is a shorthand for $(1/n!)\left(-\hat{\boldsymbol{\alpha}}_\lambda \cdot \hat{\nabla}\right)^n R_0(\mathbf{d}_0)$; and so the c_n are also implicitly dependent on the play parameters.

With the relatively coarse mesh used for plotting Fig. 8.1, all curves appear smooth. In finer detail, though, the Count-Dependent play curve has subtle kinks arising from the numerous discontinuous changes of the play parameters with true count. This effect is illustrated in Fig. 8.3, comparing the returns for six decks from Optimal Basic and Count-Dependent play but magnified by plotting the "differential" $\left[R_0\left(\langle \mathbf{d} \rangle_\gamma\right) - R\left(\mathbf{d}_0\right)\right]/\gamma$ over a narrow range of small true counts. Although the plot is itself computed with a coarse mesh, the kinks are evident. That at small positive true count is due to the change from $s = 17$ to $s = 16$ for upcard 10, while that at small negative true count corresponds to the change from $s = 12$ to $s = 13$ for upcard 4; Table 5.2 shows these shifts occurring at comparably small true count values. Similar changes in stand parameter occur for single deck, but there are also a number of less dramatic kinks corresponding to shifts in the doubling specifications at even smaller true count values. The visual evidence that the two "differential" curves have the same intercept with the vertical axis confirms that both return functions have the same slope, and not just the same value, at zero true count.

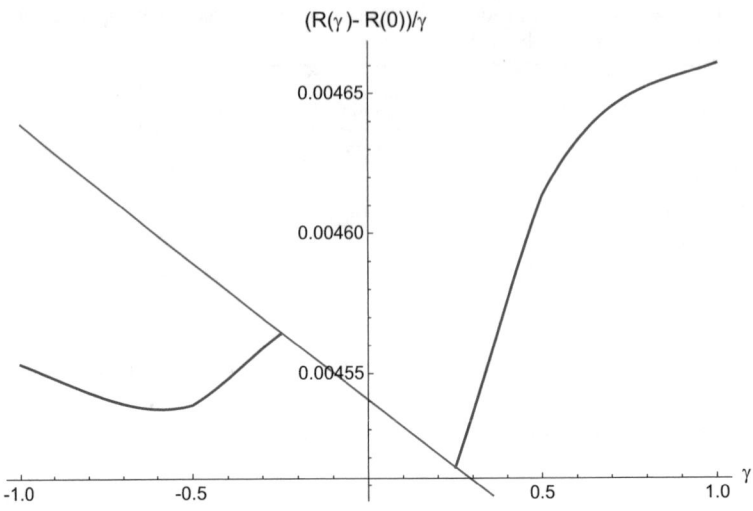

Fig. 8.3 Count-Dependent (kinked curve, blue) and Optimal Basic (smooth curve, red) first differential returns at small values of true count; 6 decks

8.3   Optimizing the Counting Vectors

The task now is to optimize Player's performance with respect to his choice of counting vectors. But since these are fixed throughout dealing the shoe, the quantity to be maximized is Player's cash return per round, averaged over the shoe. Call this Player's "yield" and denote it Y. It is given by averaging the conditional expected return of a hand over the probability distribution of $\hat{\gamma}$, weighted by the bet size B for that hand, and further averaged over all depths f of the shoe up to the penetration F:

$$Y \equiv \frac{1}{F} \int_0^F df \int d\hat{\gamma}\, \hat{p}(\hat{\gamma})\, B(\hat{\gamma}) \left\langle R_f(\mathbf{d}) \right\rangle_{\hat{\gamma}}$$

(8.24)

$$\equiv \left\langle\!\!\left\langle B(\hat{\boldsymbol{\gamma}}) \left\langle R_f(\mathbf{d}) \right\rangle_{\hat{\gamma}} \right\rangle\!\!\right\rangle,$$

where

$$\hat{p}(\hat{\gamma}) \equiv \left(\frac{1}{\sqrt{2\pi}\, \Delta} \right)^{\Lambda} \exp\left(-\frac{|\hat{\gamma}|^2}{2\,\Delta^2} \right).$$

(8.25)

To maximize Y, look for its stationary points, i.e., the zeros of its first derivatives with respect to α_λ. But because Y has been re-expressed entirely in terms of the orthonormal vectors $\hat{\boldsymbol{\alpha}}_\lambda$ which are defined as functions of the α_λ, it suffices to find the stationary points of Y with respect to the $\hat{\boldsymbol{\alpha}}_\lambda$, subject to the orthonormality conditions of Eq. (8.9). The constraints are most efficiently handled by use of Lagrange multipliers, here denoted L. The equation for the stationary points is

$$
0 = \frac{\partial}{\partial \hat{\boldsymbol{\alpha}}_\lambda} \left[Y + \sum_\lambda L_\lambda \hat{\boldsymbol{\alpha}}_\lambda \cdot \hat{\boldsymbol{d}} + \frac{1}{2} \sum_{\lambda,\lambda'} L_{\lambda,\lambda'} \left(\delta(\lambda,\lambda') - \hat{\boldsymbol{\alpha}}_\lambda \cdot \hat{\boldsymbol{\alpha}}_{\lambda'} \right) \right]
$$

$$
= \partial Y / \partial \hat{\boldsymbol{\alpha}}_\lambda + L_\lambda \hat{\boldsymbol{d}} - \sum_{\lambda'} L_{\lambda,\lambda'} \hat{\boldsymbol{\alpha}}_{\lambda'};
$$

(8.26)

while from Eq. (8.24), with (8.20) and (8.23),

$$
\frac{\partial Y}{\partial \hat{\boldsymbol{\alpha}}_\lambda} = \left\langle\!\!\left\langle B \left(-\hat{\gamma}_\lambda \hat{\nabla} - \Delta^2 \hat{\boldsymbol{\alpha}}_\lambda \cdot \hat{\nabla}\hat{\nabla} \right) \langle R_f(\mathbf{d}) \rangle_{\hat{\gamma}} \right\rangle\!\!\right\rangle .
$$

(8.27)

Equations (8.26) and (8.27) together (along with Eq. (8.9) to eliminate the Lagrange parameters) are sufficient in principle to determine the optimal $\hat{\boldsymbol{\alpha}}_\lambda$. In particular, since $\hat{\boldsymbol{d}} \cdot \hat{\nabla} = 0$, the inner product of Eq. (8.26) with $\hat{\boldsymbol{d}}$ shows that $L_\lambda = 0$. But since $\langle R_f(\mathbf{d}) \rangle_{\hat{\gamma}}$ is a nonlinear function of $\hat{\boldsymbol{\alpha}}$, no explicit expression for $\hat{\boldsymbol{\alpha}}_\lambda$ is possible in general.

8.4 Optimizing the Counting Vectors: Many-Cards Limit

The analysis to this point has not restricted the size of the pack in any way. But recognize that the number of cards in the pack is large, so that $\Delta^2 \equiv f/52 D (1-f)$ is a small parameter, less than 0.02 for $D \geq 4$ and $f \leq 0.8$. Since the Hermite expansion, Eq. (8.19), when substituted in Eq. (8.24), represents a *de facto* series expansion in Δ^2, the many-cards limit corresponds to keeping just the zeroth and first order terms in the expansion. In this case, the yield reduces to just

$$
Y = \left\langle\!\!\left\langle B \left(1 - \sum_\lambda \hat{\gamma}_\lambda \hat{\boldsymbol{\alpha}}_\lambda \cdot \hat{\nabla} \right) R_0(\mathbf{d}_0) \right\rangle\!\!\right\rangle ,
$$

(8.28)

so that

$$
\frac{\partial Y}{\partial \hat{\boldsymbol{\alpha}}_\lambda} = - \langle\!\langle B \gamma_\lambda \rangle\!\rangle \hat{\nabla} R_0(\mathbf{d}_0),
$$

(8.29)

which can then be substituted into Eq. (8.26).

The resulting equation is now linear and can be solved exactly: one counting vector, $\lambda = \lambda^*$, is given by

$$
\hat{\boldsymbol{\alpha}}^* \equiv \hat{\boldsymbol{\alpha}}_{\lambda^*} = -\hat{\nabla} R (\mathbf{d}_0) / \left| \hat{\nabla} R (\mathbf{d}_0) \right| \equiv \mathbf{v}_b,
$$

(8.30)

while the other $\Lambda - 1$ vectors are orthogonal to it. Also, for $\lambda \neq \lambda^*$, the solution requires $\langle\langle B\, \gamma_\lambda\rangle\rangle = 0$, implying that B cannot depend on any of these other counts. Thus the optimal counting vector in this many-cards approximation is aligned opposite to the gradient of the expected return function, irrespective of the form of B and for all depths. Also, Y is maximal if and only if the bet size and the conditional expected return are dependent on just the single counting vector $\mathbf{v}_b$.

In terms of the optimal vector Eq. (8.30), the expected return for any other counting vector in the many-cards approximation is just

$$\langle R_f(\mathbf{d})\rangle_\gamma = R_0 + C_b\,\hat\gamma\,\left|\,\hat\nabla R_0\,\right|, \tag{8.31}$$

where C_b is the "betting correlation" between the given counting vector and the optimal one, $C_b \equiv \hat{\boldsymbol\alpha}\, \cdot \mathbf{v}_b$. Clearly $C_b \leq 1$, with $C_b = 1$ for the optimum; the increment to the return from counting is directly proportional to C_b.

With the expected return given by Eq. (8.31), the yield expression (8.28) becomes

$$Y = \left\langle\!\!\left\langle B\left(R_0 + C_b\,\hat\gamma\,\left|\,\hat\nabla R_0\,\right|\right)\right\rangle\!\!\right\rangle \equiv Y_0 + C_b\,Y_1. \tag{8.32}$$

The first term of Eq. (8.32) is clearly the Basic Strategy result (first hand after a shuffle) with a bet equal to $\langle\langle B\rangle\rangle$; as already seen, this term is negative (except for $D = 1$). The second term is the increment from card counting, which is positive when B increases with $\hat\gamma$. Besides the factor of C_b, it is also proportional to the *rate* of bet variation: since $\langle\langle B\,\hat\gamma\rangle\rangle = \langle\langle \Delta^2\, d\, B/d\,\hat\gamma\rangle\rangle$ provided $B\,(\hat\gamma)$ is a differentiable function, a steep increase in B with $\hat\gamma$ enhances this term. Conversely, counting does not improve the yield if the bet size is fixed and not adjusted with the true count. As a consequence, if Dealer shuffles frequently to discourage Player's card counting, thereby reducing Δ^2, Player can counterattack only by raising the rate of increase of his bet. This was discussed in Sect. 4.1.

The maximization of yield with respect to the counting vector could also be carried out to the next higher order in $1/52D$; this would indicate, for example, whether the optimal counting vector changes with number of decks. Although the expressions become cumbersome, this change is found to be orthogonal to the lowest order term, so that it makes no contribution to the yield to this next order. Thus any dependence of α^* on number of decks contributes only at order $(1/52D)^2$, and hence is negligible.

8.5 Computation of the Derivatives of the Expected Return

The optimal counting vector and the maximized expected return and yield depend on the first and second derivatives of R_0. These cannot be evaluated analytically, but instead require numerical methods based on those of Sect. 8.2: systematically increment the likelihoods $\mathbf{d}$ by small amounts, and extract the first and second

Table 8.2 Gradient of return
and optimal counting vector

i	$\hat{\nabla}_i R_0$	$v_b(i)$	$\alpha(i) = v_b(i)/\hat{d}(i)$
1	+0.08493	−0.3561	−1.2839
2	−0.05419	+0.2272	+0.8191
3	−0.06202	+0.2600	+0.9376
4	−0.08015	+0.3361	+1.2117
5	−0.10037	+0.4209	+1.5174
6	−0.06452	+0.2705	+0.9753
7	−0.03804	+0.1595	+0.5751
8	+0.00400	−0.0168	−0.0604
9	+0.02747	−0.1152	−0.4153
10	+0.14145	−0.5931	−1.0692

finite differences in the expected return. The result for the first derivative is listed in Table 8.2. The (un-normalized) optimal counting vector, in the last column, is just that displayed in Table 3.2; it is consistent with Griffin's result (Griffin (1999), p. 71) for the shifts in return from removing one card from a single deck.

Other useful results are that $\left|\hat{\nabla} R_0\right| = 0.2385$ and $\hat{\nabla}^2 R_0 = -0.6527$.

Furthermore, it can be shown analytically that $\left|\hat{\nabla} R_0\right| = \partial R_0\left(\langle \mathbf{d}\rangle_\gamma\right)/\partial \hat{\gamma}\,\big|_{\gamma=0}$. The right-hand side of this relationship, from the polynomial fit of Table 8.1, computes to $52 \times 0.00453 = 0.2356$, comfortably close to the separately computed result for the left-hand side.

Denote as γ_0 the value of γ for which the return (when $D > 1$) crosses over between negative and positive: to lowest order in Δ^2,

$$\langle\langle R_f(\mathbf{d})\rangle\rangle_\gamma = |R_0|\left(-1 + \gamma/\gamma_0\right), \quad \gamma_0 \equiv 52\,|R_0|\,/C_b\,|\hat{\nabla} R_0|. \tag{8.33}$$

The expected return becomes positive, and to this order increases linearly, as γ exceeds γ_0. Numerical results for γ_0 are given in Table 3.4.

8.6   Unbalanced Counts

The focus thus far has been on those counting methods, termed "balanced," for which the count during a shuffle reverts to zero. An alternative methodology is to use "unbalanced" counting vectors, where the count just after a shuffle and its mean value are nonzero. The so-called Initial Running Count, or IRC in the terminology of Vancura and Fuchs (2016), is $52\,\mathbf{d}_0 \cdot \boldsymbol{\alpha}\,(1 - D)$ and the mean running count is $52\,D\,f\,\mathbf{d}_0 \cdot \boldsymbol{\alpha}/D\,(1 - f)$. But the inversion of Eq. (8.8) to return from the caret representation to the original shows that any fixed constant can be added onto each and every element of a balanced counting vector, thereby unbalancing it, without changing its correlation C_b or its yield. Thus for each balanced counting vector there exists a continuum of unbalanced ones with identical performance; and, conversely, every unbalanced counting vector performs identically to some balanced one. A rationale for considering unbalanced vectors is that one with simple integer elements and a favorable C_b can be chosen, whose balanced counterpart has non-integer elements and so may be impractical to use.

A disadvantage, on the other hand, is that immediately following a shuffle the running count for the unbalanced vector (its IRC) is not zero; this quantity would need to be subtracted off from the running count when forming the true count so as to accurately gauge the current expected return. Advocates of unbalanced vectors instead recommend a work-around. They introduce two specific values of the running count: one is the so-called pivot count, at which the running and true counts are equal, given by $\gamma_p = 52\,\mathbf{d}_0 \cdot \boldsymbol{\alpha}$. The other is the so-called key count, at which the expected return is zero. If the cross-over true count is roughly +1, independent of the number of decks or the penetration, then some algebra shows that the key count is $\gamma_k = \gamma_p + (1 - \gamma_p) D (1 - f)$. Although, strictly, the key count is dependent on the depth, mirroring the conversion between running count and true count, unbalanced count advocates replace f by a constant, typically around 0.6–0.75. Vancura and Fuchs (2016, p.75) put forward an empirical key count dependence on deck number of roughly $\gamma_k \approx (20 - 7D)/6$, together with a pivot of +4. Then they recommend linearly interpolating the running count between the pivot and the key to arrive at an approximate expected return, and a bet size.

Two concerns with this procedure can be raised: first, that it doesn't save any mental effort, since the true count conversion is replaced instead by the running count interpolation; and, second, that the resulting bet size is only approximate (except near the pivot), whereas optimal bet sizing is much more significant for Player's performance than is a small improvement in expected return via C_b. Although unbalanced counts are of some popularity with serious Players, they cannot be close to optimal.

Appendix 1: Asymptotic Distribution of Card Likelihoods

Equation (8.1), exhibited at the start of this chapter, should be recognizable to those familiar with discrete probabilities; but its asymptotic limit Eq. (8.2) may be less familiar, so the derivation is worth outlining. For compactness we use d_j^0 and d_j here rather than $d_0(j)$ and $d(j)$ as in the text body; and we set $d_j \equiv d_j^0 \left(1 + \varepsilon_j\right)$.

If the number of cards of each value is considered large, $v_j \gg 1$, then each factorial in Eq. (8.1) can be replaced by its asymptotic Stirling approximation, $n! \approx \sqrt{2\pi n}\,(n/e)^n$; Stirling is within 2% of correct even for n as small as 4, and the error diminishes rapidly as n increases. After considerable manipulation, Eq. (8.1) can then be written as the distribution of likelihoods $\mathbf{d}$,

$$p\left(\hat{M}\mathbf{d}\right) = \delta\left(\sum_j d_j^0 \varepsilon_j, 0\right)\sqrt{\frac{2\pi(52D)f\,\tilde{f}}{\prod_j\left[2\pi\,\nu_j\left(\tilde{f}+\tilde{f}\varepsilon_j\right)\left(f-\tilde{f}\varepsilon_j\right)\right]}}$$

$$\times\exp\left[\sum_j \nu_j\,\Phi\left(\varepsilon_j\right)\right],$$

(8.34)

with the further notation $\tilde{f} \equiv 1 - f$ and

$$\Phi(\varepsilon) \equiv \ln\left[\frac{f^f(\tilde{f})^{\tilde{f}}}{\left(\tilde{f}+\tilde{f}\varepsilon\right)^{\tilde{f}+\tilde{f}\varepsilon}\left(f-\tilde{f}\varepsilon\right)^{f-\tilde{f}\varepsilon}}\right].$$

(8.35)

In the asymptotic limit, the otherwise discrete variables $\mathbf{d}$ become quasi-continuous: the expected value of a function of $\mathbf{d}$, which in the discrete case sums the function over the distribution $p\left(\hat{M}\,\mathbf{d}\right)$, becomes an integration over $\mathbf{d}$. Also, the Kronecker delta converts to a Dirac delta. Furthermore, within that integration the exponential part of Eq. (8.34) is mostly very small, since ν_j is very large and Φ can be shown to be non-positive. Thus it is asymptotically valid for each d_j integral to focus on the integration range near $\varepsilon_j = 0$, where $\Phi(0) = 0$: we replace Φ and the square root in Eq. (8.34) by the lowest non-vanishing term in their series about $\varepsilon_j = 0$. In particular, $\Phi\left(\varepsilon_j\right) \approx -\tilde{f}\,\varepsilon_j^2\big/\,2f$. This is a standard mathematical procedure for taking the asymptotic limit, and usually called the "Method of Stationary Phase."

To complete the derivation, rescaling is needed for the conversion of the ten discrete sums to continuous integrations (over the range extended to $-\infty < \mathbf{d} < +\infty$), and from the Kronecker to Dirac delta functions. Also, bring back the parameter $\Delta \equiv (f/52D(1-f))^{1/2}$ and restore the original variables via the reverse substitution $\varepsilon_j = \left(d_j - d_j^0\right)/d_j^0$. Then the asymptotic limit of Eq. (8.34) becomes just the form (8.2), QED.

Appendix 2: Eigenmodes

In matrix algebra, an $n \times n$-dimensional real symmetric matrix $M_{i,j} = M_{j,i}$ is said to have eigenvectors e_i^μ, where $1 \leq \mu \leq n$, if $\sum_j M_{i,j}e_j^\mu = m_\mu e_i^\mu$ is satisfied for each μ; m_μ is termed the corresponding eigenvalue. If the eigenvectors are also chosen to be orthonormal, $\sum_{j=1}^n e_j^\mu e_j^\nu = \delta(\mu, \nu)$, then the matrix $\mathbf{M}$ can be represented in terms of its eigenmodes as $M_{i,j} = \sum_{\mu=1}^n m_\mu e_i^\mu e_j^\mu$. For a fuller exposition see, e.g., Shores (2007).

Chapter 9
Bet Strategies

9.1 Risk and Capitalization

9.1.1 Risk in a Game with Fixed Return

Blackjack is a game in which the expected return on a given round can differ from that of the previous round because the pack composition has changed; as a result of a differing return, the bet should change correspondingly. A proper analysis of risk in such a game must account for these fluctuations. Nevertheless, analysis of a simplified game in which the return and bet stay fixed is a useful warm-up exercise, in terms of both the mathematical tools employed and some characteristics of the results. The reasoning here, as in much of this Chapter, is based on and generalizes that of Werthamer (2005).

Thus, consider playing a model game consisting of a sequence of rounds, on each of which a fixed amount B is bet. Every round has an outcome in which Player wins an amount equal to his bet with probability $\xi(+1)$, loses his bet with probability $\xi(-1)$, and ties (i.e., no money is exchanged) with probability $\xi(0)$. (Ties are included because they occur in Blackjack, as seen in Chap. 2, about 10% of the time; but for now overlook the different payoffs from Player blackjack, doubling and splitting). Obviously, since these three possibilities exhaust all outcomes, $\sum_{\omega} \xi(\omega) = 1$, where $\omega = \pm 1, 0$. In this game, the expected return is $R = \sum_{\omega} \omega \, \xi(\omega)$; and the variance of the return is $\sigma^2 = \sum_{\omega} (\omega - R)^2 \xi(\omega) = 1 - \xi(0) - R^2$.

Assume that Player begins this model game with trip capital C_0, taken for convenience to be an integer multiple of B. The "coverage" is the dimensionless ratio $c_0 \equiv C_0/B$. If Player then plays N rounds where $N < c_0$, he cannot lose all his money even if he loses every round. In that case it is easy to show (later, generalized as Eq. (9.9)) that Player's expected capital becomes $\langle C \rangle_N = C_0 + N B R$, with variance $\langle C^2 \rangle_N - \langle C \rangle_N^2 = N B^2 \sigma^2$. His expected capital has grown or shrunk, directly proportional to N, depending on whether R is positive or

© Springer International Publishing AG, part of Springer Nature 2018
N. R. Werthamer, *Risk and Reward*, https://doi.org/10.1007/978-3-319-91385-8_9

negative. Furthermore, his capital is Gaussian distributed (generalized as Eq. (9.13)) such that the mean and variance conform to these expressions.

The more realistic case, $N \geq c_0$, includes situations where Player's trip capital is not big enough, and/or his bet size is too large, and/or he plays too many rounds, to cover every possible losing streak. He may then, with some non-vanishing probability, lose his entire stake and be forced to stop playing. This event is traditionally called "ruin." Deriving how Player's capital evolves, taking account of the ruin possibility, is mathematically non-trivial.

Begin that analysis with the standard trinomial expression for $p_N(n)$, the probability of N rounds resulting in a total of n_+ wins, n_- losses and n_0 ties:

$$p_N(n) = N! \prod_\omega \left(\xi(\omega)^{n_\omega} / n_\omega! \right). \tag{9.1}$$

Furthermore, we assert that the conditional probability $\hat{p}_N(n)$ of that result occurring without ruin (i.e., with "survival") is

$$\hat{p}_N(n) = \left(1 - \prod_\omega \frac{n_\omega!}{(n_\omega + \omega\, c_0)\,!} \right) p_N(n). \tag{9.2}$$

This expression holds for all n_- by interpreting $n_-!/(n_- - c_0)! = 0$ for $n_- < c_0$. Expression (9.2) can be deduced only with difficulty, but it can readily be proved by induction: if true for $\hat{p}_N$, then it can be shown to be true as well for $\hat{p}_{N+1}$, due to the basic relationship

$$\hat{p}_{N+1}(n) = \sum_\omega \hat{p}_N(n - \omega)\, \xi(\omega). \tag{9.3}$$

More useful for further analysis are the probabilities $\hat{\rho}_N(C)$ of Player's capital being C after N rounds, conditional on survival, given by

$$\hat{\rho}_N(C) \equiv \sum_n \delta(N, \sum_\omega n_\omega) \delta(C - C_0, B \sum_\omega \omega n_\omega)\, \hat{p}_N(n), \tag{9.4}$$

and its ruin-free analog, $\rho_N(C)$. The former satisfies a recursion like Eq. (9.3),

$$\hat{\rho}_{N+1}(C) = \Theta(C) \sum_\omega \hat{\rho}_N(C - \omega B)\xi(\omega), \tag{9.5}$$

using the step function $\Theta(C) \equiv 1,\ C > 0;\ \equiv 0,\ C \leq 0$. These quantities are convenient when deriving the probability L_N of ruin on the Nth round: since ruin occurs when a Player with capital B bets it all and loses, so that $L_N = \hat{\rho}_{N-1}(B)\,\xi(-1)$, substitution of Eqs. (9.1)–(9.4) and some manipulation shows that

$$L_N = \Theta(N - c_0)\,(c_0/N)\,\rho_N(0). \tag{9.6}$$

A result equivalent to Eq. (9.6) is quoted by Epstein (2009), p. 65, Eq. 3.8 and attributed to Lagrange.

Conservation of probability can be confirmed using Eq. (9.5): since

$$\sum_C \hat{\rho}_N(C) = \sum_C \hat{\rho}_{N-1}(C) - L_N, \tag{9.7}$$

hence iterating downward $N - c_0$ times gives

$$\sum_C \hat{\rho}_N(C) + \sum_{\mu=c_0}^{N} L_\mu = 1; \tag{9.8}$$

in words, the probability of survival plus the cumulative probability of ruin is unity. Furthermore, a similar approach shows that the expected capital after N rounds is

$$\langle C \rangle_N = \sum_C C \, \hat{\rho}_N(C) = \sum_C C \sum_\omega \hat{\rho}_{N-1}(C - \omega B)\xi(\omega)$$

$$= \sum_{C'} \sum_\omega (C' + \omega B)\hat{\rho}_{N-1}(C')\xi(\omega)$$

$$= \langle C \rangle_{N-1} + B R \left(1 - \sum_{\mu=c_0}^{N-1} L_\mu \right) \tag{9.9}$$

$$= C_0 + B R \left(N - \sum_{\mu=c_0}^{N} (N - \mu) L_\mu \right).$$

When there is no possibility of ruin, so that $L_\mu = 0$ for all μ, Eq. (9.9) proves the assertion regarding the expected capital made earlier.

The results so far have all been exact. But their simplicity is deceptive: after substituting Eq. (9.6), and its predecessors Eqs. (9.4) and (9.1), into Eq. (9.9), no further progress can be made towards a closed expression. Instead it is necessary to take the limit where the number of wins, losses, and ties becomes large: correspondingly, also where $N \gg c_0$. This asymptotic limit allows a closed expression for the full distribution of capital and not just its low moments, namely the risk of ruin and the expected return.

Begin the analysis by introducing a Fourier representation for the capital distribution and the ruin probability,

$$F_N(\phi) \equiv \int dC \, e^{i(c-c_0)\phi} \rho_N(C), \quad L_N(\phi) \equiv e^{-ic_0\phi} L_N,$$

with the reduced variables $c \equiv C/B$ and $c_0 \equiv C_0/B$. Then the recursion relation on $\hat{\rho}_N(C)$, Eq. (9.5), translates into a recursion on $F_N(\phi)$,

$$F_N(\phi) = v(\phi) F_{N-1}(\phi) - (v(\phi) - 1) \sum_{\mu=c_0}^{N-1} L_\mu(\phi), \tag{9.10}$$

where $v(\phi) \equiv \sum_\omega e^{i\omega\phi}\xi(\omega)$. Iterating this recursion another $N - c_0 - 1$ times, and using $F_{c_0} = v(\phi)^{c_0}$, leads to

$$F_N(\phi) = v(\phi)^N - \sum_{\mu=c_0}^{N} v(\phi)^{N-\mu} - 1 L_\mu(\phi); \tag{9.11}$$

the survival portion of $F_N(\phi)$ is just

$$\hat{F}_N(\phi) = v(\phi)^N - \sum_{\mu=c_0}^{N} v(\phi)^{N-\mu} L_\mu(\phi). \tag{9.12}$$

Furthermore, the definition of $v(\phi)$ facilitates extending the ω summation to include blackjack ($\omega = +3/2$) and double/split ($\omega = \pm 2$).

Since taking the asymptotic limit of Eq. (9.12) is a lengthy derivation, it is relegated to Appendix 1; but the result for the distribution of capital with survival is

$$\hat{\rho}_N(C) \approx \frac{1}{\sqrt{2\pi N B^2 \sigma^2}} \left\{ \exp\left(-\frac{(C - C_0 - NBR)^2}{2NB^2\sigma^2} \right) \right.$$
$$\left. - \exp\left(-\frac{2C_0 R}{B\sigma^2} - \frac{(C + C_0 - NBR)^2}{2NB^2\sigma^2} \right) \right\}. \tag{9.13}$$

The expected capital and risk of ruin, then, are

$$\langle C \rangle_N \approx \int_0^\infty C \hat{\rho}_N(C)\, dC$$

$$\approx (C_0 + NBR) - \exp\left(-\frac{C_0 \bar{R}}{B\sigma^2} \right) \sum_{\pm} \exp\left(\pm\frac{C_0 R}{B\sigma^2} \right)$$
$$\times (NBR \pm C_0)\, \mathrm{erf}\left(\frac{C_0 \pm NBR}{NB^2\sigma^2} \right) \tag{9.14}$$

and

$$W_N \equiv 1 - \int_0^\infty \hat{\rho}_N(C)\,dC$$
$$\approx \exp\left(-\frac{C_0 \bar{R}}{B\sigma^2} \right) \sum_{\pm} \exp\left(\pm\frac{C_0 R}{B\sigma^2} \right) \mathrm{erf}\left(\frac{C_0 \pm NBR}{NB^2\sigma^2} \right), \tag{9.15}$$

in terms of the error function (sometimes called the complementary normal integral)

$$\mathrm{erf}(x) \equiv \frac{1}{\sqrt{2\pi}} \int_x^\infty dy \, \exp\left(-\frac{y^2}{2} \right); \tag{9.16}$$

a direct integration shows that erf (0) = 1 / 2, and hence that erf($-\infty$) = 1. Further, the quantity $\bar{R} \equiv -R + \sigma^2 \ln\left((1 + R/\sigma^2)/(1 - R/\sigma^2)\right)$. Equations (9.13)–(9.15) hold for both positive and negative R.

In the very large N limit, $N \gg R^{-2}$, Eq. (9.15) reduces to just $W_N \approx \exp\left[(C_0/B) \ln\left((1 - R/\sigma^2)/(1 + R/\sigma^2)\right)\right]$. The similar but not identical expression, $W_N = \exp\left[(C_0/B\sigma) \ln((1 - R/\sigma)/(1 + R/\sigma))\right]$, with σ^2 as the exact variance, is quoted without proof both by Carlson (2017), p. 156 and by Schlesinger (2005), p. 112. It appears still earlier as Eq. 5 of Sileo (1992), who apparently misuses a model due to Griffin (1999), pp. 141–142. The expression here, however, does agree with Epstein (2009), pp. 58–59, Eq. 3.2ff. For the remainder of this chapter, assume the return is small, $R \ll 1$ so that this discrepancy disappears and $\bar{R} \simeq \hat{R}$.

Comparison of Eqs. (9.14) and (9.15) shows that $(\langle C \rangle_N - C_0)/ N B R$ is just the probability of survival at N but with a correction term that removes random walks into negative capital prior to N. Both Eqs. (9.14) and (9.15) are straightforward to evaluate numerically and graph, although the plots are deferred to the end of the next section; however, the correction term in Eq. (9.15) does not have a major qualitative effect.

A three-dimensional plot of the capital distribution and its low moments, Eqs. (9.13)–(9.15), is displayed in Fig. 9.1. For small numbers of hands the distribution is sharply peaked at a capital ratio of unity; as the number of hands increases the distribution spreads out. But because of the possibility of ruin, the

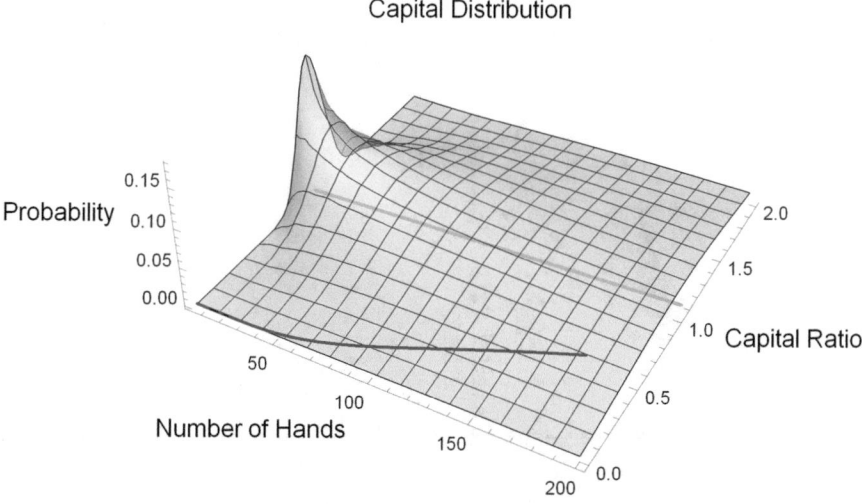

Fig. 9.1 Distribution of capital, scaled to its initial value, vs. number of hands played. Shown is the plot for initial capital of 20 unit bets and expected return of 0.01. To interactively select other values for these parameters, browse to the Wolfram Cloud website www.wolfr.am/BlackjackScience and select the appropriate value. This figure is copyright by N. Richard Werthamer and used by permission

distribution is always constrained to zero at zero capital. Also shown are plots of the ruin probability (red, in the plane of probability vs. number of hands) and of the expected capital ratio (green, in the plane of capital ratio vs. number of hands). The rising probability of ruin with increasing number of hands exactly compensates the decreasing area under the capital ratio surface, i.e., the decreasing probability of survival.

9.1.2 Optimal Betting When Return Fluctuates: Expected Capital and Risk

The instructive but simplified warm-ups of the previous section facilitate taking on the real game of Blackjack, where the return R varies from round to round. Section 8.1 has already demonstrated that R is distributed about a mean, $\langle R \rangle$, with a width that increases with the fraction of the pack dealt out. For a game with multiple decks, $\langle R \rangle$ is negative. That part of the distribution where $R > 0$ offers the opportunity to increase the bet size, so that $B \rightarrow B(R)$. These features are illustrated in Fig. 9.2.

Because each round has a different return and bet size, label the rounds by an index, μ, and generalize the quantities of the previous subsection to R_μ, B_μ, $v_\mu(\phi)$.

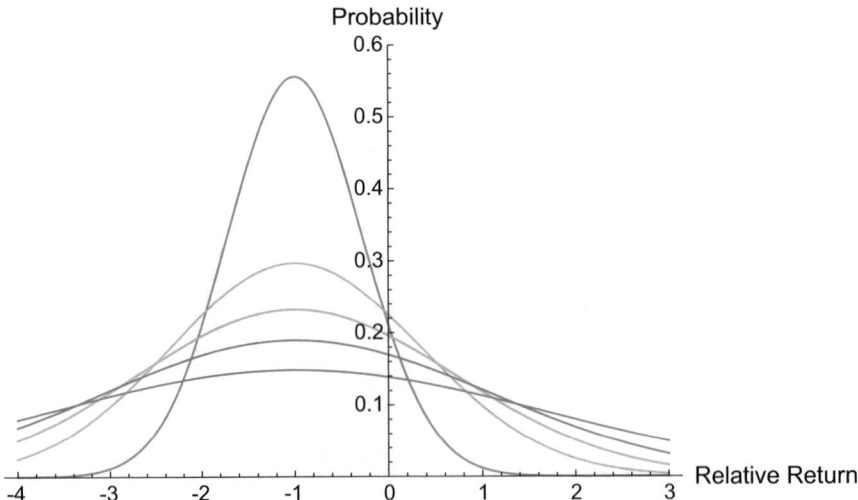

Fig. 9.2 Distribution of return, scaled to its magnitude at zero count, for various depths, 4 decks. The plot shows that as depth increases from 0.05 (blue) to 0.2 (yellow) to 0.4 (green) to 0.6 (red) to 0.8, the distribution widens from its narrow peak around -1 and flattens; the spreading is more pronounced for one deck than for larger sizes. To interactively select other values for these parameters, browse to the Wolfram Cloud website www.wolfr.am/BlackjackScience and select the appropriate value. This figure is copyright by N. Richard Werthamer and used by permission

The bet B_μ is some integer multiple of a base bet B_0. Then the probability distribution $\rho_N(C)$, at first in the absence of ruin, is naturally expressed in the Fourier representation as

$$\rho_N(C) = \frac{B_0}{2\pi} \int\limits_{-\pi/B_0}^{\pi/B_0} d\phi \, \exp\left(-i\,(C - C_0)\,\phi + \sum_{\mu=1}^{N} \ln v_\mu\,(\phi)\right). \qquad (9.17)$$

Once again, temporarily disregard the distinctive payoffs to hands that are doubled, split or blackjack.

For a sufficiently long sequence of rounds, through multiple reshuffles of the shoe, the sum over μ tends to its ensemble average. Then, in Eq. (9.17),

$$\sum_{\mu=1}^{N} \ln v_\mu(\phi) \approx N\langle\langle v\,(\phi)\rangle\rangle \approx \left\langle\!\left\langle \hat{N}\left(i\,B\,\hat{R}\,\phi - B^2\phi^2/2\right)\right\rangle\!\right\rangle; \qquad (9.18)$$

the second equivalence in Eq. (9.18) takes the small-argument expansion of $v\,(\phi)$, appropriate for the asymptotic, large N limit. The ϕ integral, just as in the derivation of Eq. (9.13), then gives

$$\rho_N(C) = \frac{B_0}{\left(2\pi\left\langle\!\left\langle\hat{N}\,B^2\right\rangle\!\right\rangle\right)^{1/2}}\,\exp\left[-\frac{(C - C_0 - N\langle\langle B\,R\rangle\rangle)^2}{2\left\langle\!\left\langle\hat{N}\,B^2\right\rangle\!\right\rangle}\right]. \qquad (9.19)$$

Comparison of Eqs. (9.13) and (9.19) shows that when the return and bet size vary, the expected capital after N rounds without ruin becomes $\langle C\rangle_N = C_0 + N\langle\langle B\,R\rangle\rangle$. It grows with N to the extent that $\langle\langle B\,R\rangle\rangle > 0$, even if $\langle R\rangle < 0$. Note that $\langle\langle B\,R\rangle\rangle$ is just the (risk-free) yield per round, Y, central to Chaps. 3 and 4. Thus a goal for bet strategy is to arrive at a functional dependence of B on R for which the yield is positive, and the expected capital grows rather than shrinks with N.

The analysis of this subsection has so far neglected ruin, thus omitting the key source of risk. To include this effect, return to Eq. (9.5) and generalize it to

$$\hat{\rho}_N(C) = \Theta(C)\sum_\omega \hat{\rho}_{N-1}\,(C - \omega\,B_N)\,\xi_N(\omega). \qquad (9.20)$$

Also, the ruin probability becomes $L_N = \hat{\rho}_{N-1}\,(B_N)\,\xi_N(-1)$; and $L_N = 0$ if $N < c$, where the coverage c is now defined via

$$C_0 = \sum_{\mu=1}^{c} B_\mu \equiv c\,\langle\langle B\rangle\rangle. \qquad (9.21)$$

A related coverage parameter that will be especially useful later is $\bar{c} \equiv C_0/\langle\langle B^2\sigma^2\rangle\rangle^{1/2}$, the ratio of trip capital to what in Sect. 4.1 is called the typical bet. Here, "typical" means the root-mean-square total amount wagered on a hand, including the increased initial bets on hands with positive expected return and the additional amounts needed to cover doubles and splits.

The analysis leading to Eqs. (9.8) and (9.9) proceeds as before: probability is conserved, just as expressed by Eq. (9.8), and the expected capital becomes

$$\langle C \rangle_N = C_0 + \sum_{\mu=1}^{N} B_\mu R_\mu - \sum_{\mu=1}^{N} L_\mu \sum_{\nu=\mu+1}^{N} B_\nu R_\nu. \qquad (9.22)$$

For large N, the sum $\sum_\nu B_\nu R_\nu$ includes enough terms that it approaches its ensemble average; furthermore, it is uncorrelated from the outcome of all rounds $\nu \le \mu$. Hence

$$\langle C \rangle_N = C_0 + \langle\langle B\,R \rangle\rangle \left(N - \sum_{\mu=c}^{N} (N - \mu)\, L_\mu \right). \qquad (9.23)$$

In order to proceed further, a generalization of Eq. (9.6) is needed; it could then be combined with Eqs. (9.19) and (9.23) to derive a generalization of Eq. (9.10). To accomplish this, first re-derive Eq. (9.6) by iteration of $L_N = \hat\rho_{N-1}(B)\,\xi(-1\ )$, but instead use Eq. (9.3) and avoid Eq. (9.2). The result, in schematic form, is

$$L_N = \sum_{\text{perms}} \delta \left(-C_0,\ B \sum_\omega \xi\,(\omega) \right) \prod_\omega \xi(\omega)^{n_\omega}, \qquad (9.24)$$

where the sum is over all permutations of sequences in which the capital goes from C_0 to 0 in N rounds without reaching 0 at any intermediate round. But from Eq. (9.6), when the probability factors are the same for every hand,

$$\sum_{\text{perms}} \cdots = \frac{C}{N} N! \sum_{\mathbf{n}} \delta \left(N,\ \sum_\omega n_\omega \right) \prod_\omega \left(\frac{1}{n_\omega!} \right) \cdots . \qquad (9.25)$$

By introducing the Fourier representation for the Kronecker delta function, Eq. (9.24) can be expressed as

$$L_N = \frac{B}{2\pi} \int_{-\pi/B}^{\pi/B} d\phi\, \exp\,(i\, C_0\, \phi) \sum_{\text{perms}} \prod_\omega (\xi(\omega)\exp(i\,\omega\,B\phi))^{n_\omega}. \qquad (9.26)$$

When the returns and bets vary from round to round, Eq. (9.26) for L_N generalizes to

$$L_N = \frac{B_0}{2\pi} \int_{-\pi/B_0}^{\pi/B_0} d\phi\, \exp\,(i\, C_0\, \phi) \sum_{\text{perms}} \prod_\omega \prod_{\mu_\omega} \xi_{\mu_\omega}(\omega) \exp\,(i\,\omega\,B_{\mu_\omega}\phi),$$

$$(9.27)$$

where the products over the μ_ω, respectively, contain n_ω terms. Again as for Eq. (9.15), each product contains enough terms, when N is large, to tend to its ensemble average and is uncorrelated from the outcomes of the other sequences. Then

$$L_N \approx \frac{B_0}{2\pi} \int\limits_{-\pi/B_0}^{\pi/B_0} d\phi \, \exp\left(i \, C_0 \, \phi\right) \sum_{\text{perms}} \prod_\omega \langle\langle \xi(\omega) \exp(i \, \omega \, B \, \phi) \rangle\rangle^{n_\omega}$$

$$\approx \frac{B_0 \, C}{2\pi \, N} \int\limits_{-\pi/B_0}^{\pi/B_0} d\phi \, \exp\left(i \, C_0 \, \phi + N \ln\langle\langle v(\phi) \rangle\rangle\right).$$

(9.28)

The second line follows by using Eq. (9.25) and the argument leading to Eq. (9.13). The last step in the derivation points out that

$$N \, \ln\langle\langle v \, (\phi) \rangle\rangle \approx \left\langle\left\langle \hat{N} \left(i \, B \, \hat{R} \, \phi - B^2 \, \phi^2/2 \right) \right\rangle\right\rangle$$

$$\approx N \langle\langle \ln v(\phi) \rangle\rangle$$

(9.29)

in the asymptotic limit, using $|B \, \phi| \ll 1$ and that the variance of the yield $\ll$ the mean square bet size. Thus Eqs. (9.28) and (9.29), compared to Eqs. (9.19)–(9.21), show again that $L_N \approx \Theta(N - c) \, (c/N) \, \rho_N(0)$ as anticipated. Substituting into Eq. (9.23) leads to a form for $\langle C \rangle_N$ just like Eq. (9.14) but with the replacements $B \, R \rightarrow \langle\langle B \, R \rangle\rangle$, $c_0^2/\hat{N} \rightarrow C_0^2 / N \langle\langle B^2 \sigma^2 \rangle\rangle = \bar{c}^2/N$, and $\hat{R} \rightarrow \langle\langle B \, R \rangle\rangle C_0 / \langle\langle B^2 \, \sigma^2 \rangle\rangle \equiv q$: explicitly,

$$\frac{\langle C \rangle_N - C_0}{N \langle\langle B \, R \rangle\rangle} = 1 - \exp\left(-q\right) \sum_{\pm} \exp(\pm q) \left(1 \pm \frac{\bar{c}^2/N}{q}\right) \text{erf}\left(\frac{(\bar{c}^2/N) \pm q}{(\bar{c}^2/N)^{1/2}}\right)$$

$$\equiv \Psi_N.$$

(9.30)

Ψ_N, of course, is just the yield reduction factor (YRF) discussed in Sect. 4.1. Adding plausibility to this result for $\langle C \rangle_N$ is its close parallel to the generalization, in the absence of ruin, that takes the distribution of C from Eq. (9.13) for fixed betting to Eq. (9.19) for variable betting. Equation (9.15), for the cumulative ruin probability, similarly generalizes to

$$W_N \rightarrow \exp\left(-q\right) \sum_{\pm} \exp(\pm q) \, \text{erf}\left(\frac{(\bar{c}^2/N) \pm q}{(\bar{c}^2/N)^{1/2}}\right).$$

(9.31)

Schlesinger (2005), p. 132 displays, without proof but attributed to Chris Cummings (private communication), an expression equivalent to Eq. (9.31).

9.1.3 Connections with Finance

Although expression (9.31) is relatively unfamiliar in blackjack analysis, it is similar
to a celebrated milestone in the mathematics of finance: the Black-Scholes formula
for the rational price of an option on an asset, such as a stock, whose price fluctuates.
This result, derived by Fischer Black and Myron Scholes (1973), and by Robert
Merton (1973), earned the Nobel Prize in Economics for Scholes and Merton
(Black also would have been honored, but had died before the prize was awarded).
Finding such a close connection between a game like blackjack on the one hand,
where the outcome of each round is statistically independent of all previous rounds
(i.e., a Markov stochastic process), and the stock market on the other requires the
observation, which Black, Scholes, and Merton (BSM) adopted from earlier work of
Samuelson (1965), that fluctuations of the logarithm of stock prices approximately
obey Markov statistics.

BSM take a different mathematical path to their result than here, where the
rounds of play were first regarded as discrete, arriving at expressions (9.2) and (9.9),
and then the asymptotic, or continuum, limit was taken to reach (9.10) and (9.12).
BSM instead assume the continuum limit at the start of their derivation, appeal to
the stochastic calculus of Ito (1951), and obtain their formula as the solution of
a drifting diffusion equation. Their approach is plausible since a Markov process
generates a random walk, whose continuum limit is diffusion. The connection with
the approach here can be confirmed directly by verifying that the expected capital,
Eq. (9.30), satisfies the standard diffusion equation,

$$\left(\frac{\partial}{\partial N} - \langle\langle B\,R\rangle\rangle \frac{\partial}{\partial C_0} - \frac{1}{2}\langle\langle B^2\sigma^2\rangle\rangle \frac{\partial^2}{\partial C_0^2} \right) \langle C\rangle_N = 0 \tag{9.32}$$

—as does the cumulative ruin probability, Eq. (9.33), as well—with diffusion
constant $\langle\langle B^2\sigma^2\rangle\rangle$ and drift, or bias, $\langle\langle B\,R\rangle\rangle$; the number of rounds is a proxy for
time. Skipping the discrete model in favor of the continuum could have led to
Eq. (9.30), for instance, just by solving the diffusion equation with the conditions
$\langle C\rangle_{N\to 0} \to C_0$, $\langle C\rangle_N = 0$ for $C_0 = 0$ and $N > 0$. However, since the
approach here only requires the Stirling approximation and a steepest descents
integration, Eqs. (9.30)–(9.31) can be conjectured (see below) to already be close
approximations for $N \approx 50$, not nearly so large as the true continuum limit.

An approach to deriving the Black-Scholes result based on discrete steps, as here,
was later taken by Cox, Ross, and Rubenstein (1979), CRR. Their rationale was
to circumvent the sophisticated continuum mathematics of the stochastic calculus.
They start with a simple binary model in which equity prices fluctuate either up or
down by a fixed amount (like here, but without ties), such that the log of the price
conforms to a Markov process. Then they take the limit of increasingly short time
steps and small fluctuation amplitudes. Since this asymptotic limit, applied in their
case to the value of an option, involves only the mean and variance of the Markov
process, they are able to argue that the result would also hold for any more complex

process, such as the actual stock market, provided only that the mean and variance of the model are matched exactly with those of the market. Their argument is an alternative way to derive a diffusion equation like Eq. (9.32) whose coefficients are averages over the actual process—blackjack in our application, equity markets in theirs. The CRR approach thus provides a direct proof of Eq. (9.30), alternative to the argument given here. Furthermore, subsequent simulations with the CRR model, e.g., Hull (2002), have demonstrated that the continuum limit of the discrete step model is approached quite closely in well under a hundred steps, bolstering the conjecture above.

9.1.4 Distribution of Capital

The mathematical approaches used in the theories of finance, sketched above, suggest a mathematical route to the full statistics of Player's capital with varying returns and bet sizes; this, in turn, provides an alternative derivation of W_N and $\langle C \rangle_N$. The distribution function $\rho_N(C) = W_N \delta(C) + \hat{\rho}_N(C)$ is the sum of a contribution from the probability of ruin, W_N, weighted entirely at $C = 0$, and a term $\hat{\rho}$ which satisfies a drifting diffusion equation analogous to Eq. (9.32),

$$\left(\frac{\partial}{\partial N} + \langle\langle B\,R \rangle\rangle \frac{\partial}{\partial C} - \frac{1}{2} \langle\langle B^2 \sigma^2 \rangle\rangle \frac{\partial^2}{\partial C^2} \right) \hat{\rho}_N(C) = 0. \tag{9.33}$$

The solution with the boundary condition $\hat{\rho}_N(0) = 0$ and initial condition $\hat{\rho}_0(C) = \delta(C - C_0)$ is the superposition

$$\hat{\rho}_N(C; C_0) = \frac{1}{\left(2\pi N \langle\langle B^2 \sigma^2 \rangle\rangle\right)^{1/2}} \left\{ \exp\left[-\frac{(C - C_0 - N\langle\langle B\,R \rangle\rangle)^2}{2N \langle\langle B^2 \sigma^2 \rangle\rangle} \right] \right.$$

$$\left. - \exp\left[-\frac{2\langle\langle B\,R \rangle\rangle C_0}{\langle\langle B^2 \sigma^2 \rangle\rangle} - \frac{(C + C_0 - N\langle\langle B\,R \rangle\rangle)^2}{2N \langle\langle B^2 \sigma^2 \rangle\rangle} \right] \right\}. \tag{9.34}$$

This expression, generalizing Eq. (9.13) to variable returns and bet sizes, also has the intuitively plausible convolution property,

$$\int_0^\infty \hat{\rho}_{N_2 - N_1}(C_2; C_1)\, \hat{\rho}_{N_1}(C_1; C_0)\, dC_1 = \hat{\rho}_{N_2}(C_2; C_0), \tag{9.35}$$

although the manipulations to verify the result, after substituting Eq. (9.34) into Eq. (9.35), are tedious.

The first two moments of $\hat{\rho}$ lead again to Eqs. (9.30) and (9.31). In particular, the normalization $\int_{-\infty}^\infty \rho_N(C) dC = 1$ implies that $W_N = 1 - \int_0^\infty \hat{\rho}_N(C; C_0)\, dC$; carrying out the integration, and reintroducing the original parameters, reproduces Eq. (9.31). Furthermore, the first moment $\langle C \rangle_N = \int_0^\infty C \hat{\rho}_N(C; C_0)\, dC$ similarly reproduces Eq. (9.30).

9.1.5 Properties of the Risk and Expected Capital Expressions

Having confirmed that Eqs. (9.30)–(9.31) are the appropriate generalization of
Eqs. (9.14) and (9.15) for varying returns and bets, they can now be analyzed in
more detail. The expected capital and ruin probability depend on the two parameters,
$N / \bar{c}^2$ and q, each of which can range from 0 to ∞; the expressions also depend
on the sign of $Y = \langle\langle B R \rangle\rangle$, although here a bet scheme with positive yield
is assumed. Numerical evaluations of Eqs. (9.30)–(9.31) are straightforward. The
survival probability, $1 - W_N$, and the Yield Reduction Factor, YRF, are plotted in
Fig. 9.3 on a log-log scale. These plots are a different way of looking at the closely
related curves in Fig. 9.1: but here it is recognized explicitly that only two of the
three parameters of those curves are independent.

Although the survival probability is not precisely identical to the YRF, the two
are qualitatively similar and differ by at most a factor of two. Figure 9.3 illustrates
that the two quantities each vary monotonically with q (i.e., always increase as q
increases, and vice versa), so q is called the ruin protection in Sect. 4.1. When q is
large enough (e.g., comparable to or larger than one), ruin is very unlikely and the
YRF approaches unity for all but extraordinarily large values of $N / \bar{c}^2$ (as shown
in the left-hand portion of the plot). But for q significantly less than one, the YRF
decreases steeply with increasing $N / \bar{c}^2$ (right-hand portion), especially after many
rounds such that $N \gg \bar{c}^2$. The conclusion is that ruin can be a decided risk if the
coverage is insufficient.

A way to understand these results more intuitively is to look back at Eq. (9.19) for
the probability distribution of capital, C, without ruin. The peak of the distribution
drifts linearly with increasing N to larger positive values, while the width increases
as $N^{1/2}$. Ruin becomes likely to the extent that the distribution has any significant
weight at $C = 0$, i.e., when $(C_0 + N \langle\langle B R \rangle\rangle)^2 < N \langle\langle B^2 \sigma^2 \rangle\rangle$. This condition, in
turn, is satisfied if the two conditions $q < \bar{c}/N^{1/2} < 1$ both hold, much like the
conditions (low ruin protection and low capital coverage) of the previous paragraph
(the foreground of Fig. 9.3). Ruin occurs when the distribution, despite the steady
drift of its peak to more positive values, has diffused outward to overlap $C = 0$; see
Fig. 9.4. Ruin protection affects the relation of the drift to the broadening; large ruin
protection q, as well as large coverage $\bar{c}$, keeps the overlap exponentially small and
forestalls ruin.

9.1.6 Optimal Betting When Return Fluctuates: Bet Strategy

Having derived the distribution of Player's capital, dependent on the ruin protection
and the number of rounds, the bet strategy can now be optimized. The criterion for
optimization invoked here is to find the bet function that maximizes the expected
capital for a given number of rounds, yet fixes the cumulative ruin probability
at a predetermined value. Microeconomists might instead prefer the introduction

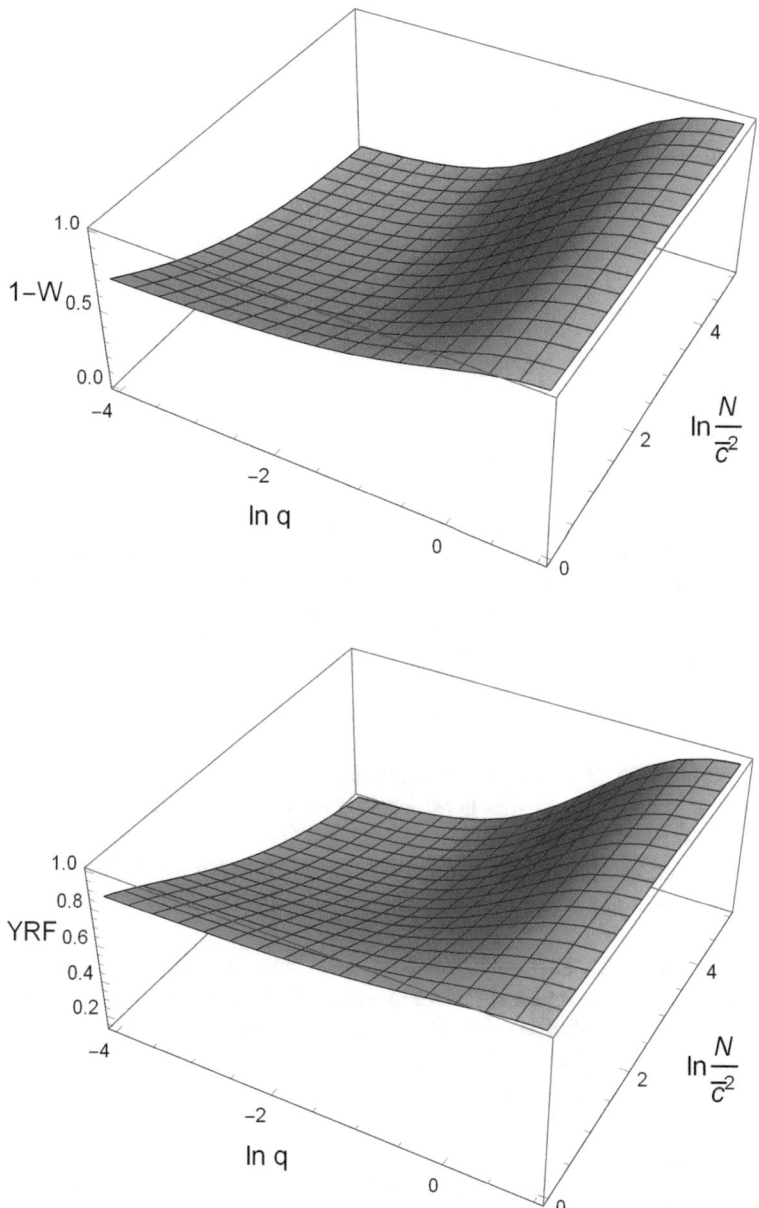

Fig. 9.3 Survival probability (upper) and Yield Reduction Factor (lower) as functions of the parameters **q** and N, on a log-log scale. To access an interactive three-dimensional rotation feature, browse to the Wolfram Cloud website www.wolfr.am/BlackjackScience, select this figure, and click-drag the cursor on the image. This figure is copyright by N. Richard Werthamer and used by permission

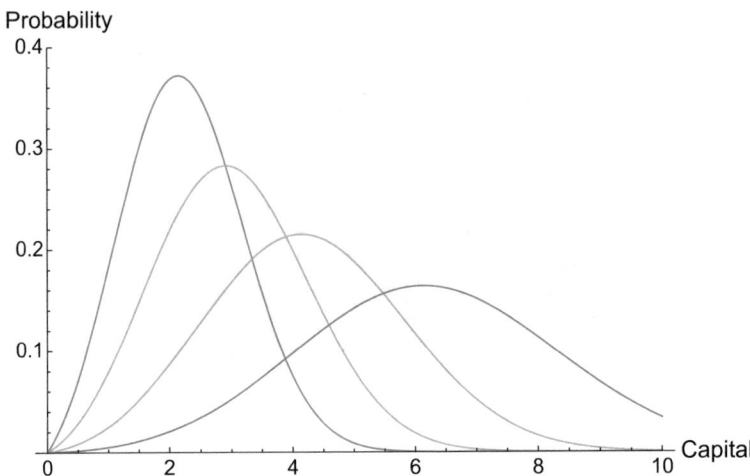

Fig. 9.4 Distribution of capital, normalized to its initial value, for various values of the number of rounds, N, and with ruin parameter q=0. The plot shows that as N increases (blue to orange to green to red) the distribution shifts, widens, and flattens—its mean and variance both increase. The possibility of ruin restricts the distribution to non-negative capital. To interactively select other values for N and q, both indexed logarithmically, browse to the Wolfram Cloud website www. wolfr.am/BlackjackScience and select the appropriate value. This figure is copyright by N. Richard Werthamer and used by permission

and maximization of a Utility Function (see, e.g., Ingersoll (1987), especially Chapter 1), but such a criterion is primarily a formal one, difficult to apply for any practical results.

Thus, functionally maximize $\langle C \rangle_N$—or more conveniently an "effective yield" $y_N \equiv (\langle C \rangle_N - C_0)/ N\, C_0 = Y\, \Psi_N/C_0$—with respect to $B(\gamma)$ while holding W_N constant.

However, $B(\gamma)$ is also constrained to lie within the range between the table minimum B_- and maximum B_+ as set by House rules. If Player elects, for whatever reason, to cap his bet size below the table maximum, then B_+ instead represents that cap. Similarly, he might also choose his base bet, B_-, above the table minimum, as long as $B_- \leq B_+$. In any event,

$$\langle\langle B\,R \rangle\rangle = \left(B_- \int_{-\infty}^{\gamma_-} + B_+ \int_{\gamma_+}^{\infty} \right) d\gamma\, \rho(\gamma)\, R(\gamma) + \int_{\gamma_-}^{\gamma_+} d\gamma\, \rho(\gamma)\, B(\gamma)\, R(\gamma) \quad (9.36a)$$

and a similar expression for $\langle\langle B^2\sigma^2 \rangle\rangle$, referenced later as Eq. (9.36b). Since $B(\gamma)$ is intuitively expected to increase monotonically with γ, define $\gamma_\pm$ via $B(\gamma_\pm) = B_\pm$. The "spread," then, is $\beta \equiv B_+/B_-$.

The maximization of y_N with respect to $B(\gamma)$ in the range $\gamma_- \leq \gamma \leq \gamma_+$—with the constraint of fixed W_N best handled by use of a Lagrange multiplier, denoted

L—leads to the functional stationary equation $\delta\,(y_N - LW_N)\big/\,\delta\,B\,(\gamma) = 0$. The differentiation at first looks complicated; but the complexity can be circumvented by recognizing, from Eqs. (9.30) and (9.31), that the expression depends on $B(\gamma)$ only via $\langle\langle BR \rangle\rangle$, whose functional derivative is proportional to $R(\gamma)$, and on $\langle\langle B^2\sigma^2 \rangle\rangle$ whose functional derivative is proportional to $B(\gamma)\sigma^2$. Thus the result of differentiation is just a linear combination of these two simple expressions, and so the stationary equation has the solution $B(\gamma) \propto R(\gamma)\big/\sigma^2$. As expected, the bet is proportional to R between a lower and upper threshold. Define a constant of proportionality s such that $b(\gamma) \equiv B(\gamma)\big/C_0 = s\,R(\gamma)\big/\sigma^2$ in this range. Then the threshold true counts satisfy $R(\gamma_\pm) = b_\pm\,\sigma_\pm^2\big/s$. A bet function of this form is consistent with an existence theorem asserted by Epstein (2009), p. 66, and referenced by Griffin (1999), p. 139. Substitution of this functional form for $b(\gamma)$ into Eqs. (9.36a) and (9.36b) leads to the expressions

$$\langle\langle b\,R \rangle\rangle = A_1 + s\,A_2, \; \langle\langle b^2\sigma^2 \rangle\rangle = A_0 + s^2\,A_2, \tag{9.37}$$

where the A coefficients are defined as weighted integrals over γ.

Although the expected return in general is a non-linear function of true count—Sect. 8.2 has shown it can be closely approximated by a cubic polynomial, although the deviation from linearity becomes significant only for large values of true count—we temporarily take the function to be linear and hence independent of depth. The distribution of returns is then Gaussian just like that of true counts; and we can convert the $A_{0,1,2}$ quantities into the integrals over R,

$$A_0 \equiv \left(b_-^2 \int\limits_{-\infty}^{R_-} + \, b_+^2 \int\limits_{R_+}^{\infty} \right) d\,R\; \sigma^2\,\rho(R),$$

$$A_1 \equiv \left(b_- \int\limits_{-\infty}^{R_-} + \, b_+ \int\limits_{R_+}^{\infty} \right) d\,R\; R\,\rho(R), \tag{9.38}$$

$$A_2 \equiv \int\limits_{R_-}^{R_+} d\,R\; \left(\frac{R}{\sigma}\right)^2 \rho(R).$$

The weight $\rho(R)$ incorporates the average of that Gaussian distribution over all depths f of the pack up to the reshuffle penetration F:

$$\rho(R) = \frac{1}{F}\int\limits_{0}^{F} df\; \frac{1}{\sqrt{2\pi}\,\Delta_R}\,\exp\left[-\frac{1}{2}\left(\frac{R - R_0}{\Delta_R}\right)^2\right], \Delta_R \equiv 52\,\Delta\,R_0. \tag{9.39}$$

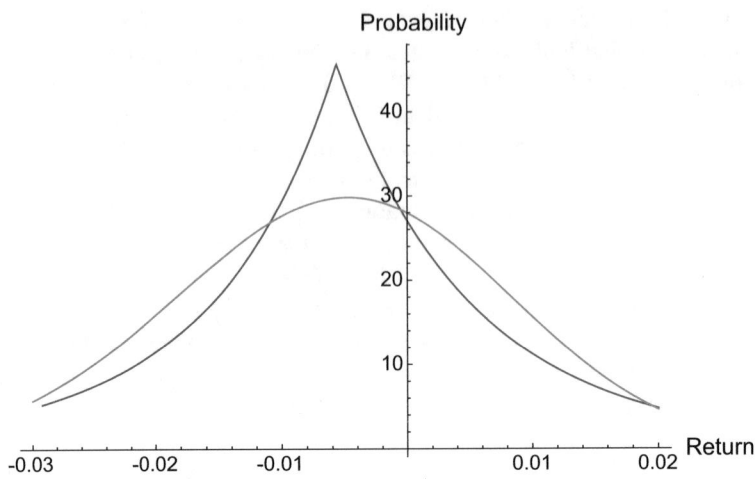

Fig. 9.5 Distributions of expected returns: at depth 0.4 (Gaussian curve, blue) and averaged over depths to a penetration 0.8 (peaked curve, red), for representative game parameters. To interactively select other values for these parameters, browse to the Wolfram Cloud website www.wolfr.am/BlackjackScience and select the appropriate value. This figure is copyright by N. Richard Werthamer and used by permission

The slope parameter s is set by the specified level of risk, W_N, which in turn is parameterized by q and $\bar{c}$. In principle it would be desirable to invert that relationship to find s as a function of W_N, and substitute into y_N. But the inversion is not possible in closed form, instead requiring computation. The need to specify the two bet threshold parameters $b_\pm$ creates additional complexity for a numerical approach.

To gain some partial insights, begin the computational work by choosing the representative set of parameters for a blackjack game adopted in Sect. 3.4: 6 decks with penetration at 80% of the shoe before reshuffle; Generic Strategy expected return of -0.005, with variance of 1.26 independent of R; and zero expected return at a true count of $+1$. With these parameters, the distribution of returns becomes

$$\rho(R) = \frac{1}{0.8} \int\limits_0^{0.8} df \, \frac{1}{\sqrt{2\pi}\,\Delta_R} \exp\left(-\frac{1}{2}\left(\frac{R+0.005}{\Delta_R}\right)^2\right),$$

(9.39a)

$$\Delta_R^2 = \frac{52(0.005)^2 f}{6(1-f)}.$$

Figure 9.5 graphs $\rho(R)$, along with its integrand at $f = 0.4$, showing how the averaging over depth distorts the shape away from a Gaussian. Selecting values for $b_\pm$ and N permits computation of both y_N and W_N vs. the slope s. An example is shown in Fig. 4.2 with the Lifetimer parameters, $b_- = 0.001$, $b_+ = 0.010$, $N = 10^6$. Each curve has an extremum: y_N has a shallow maximum while, at a smaller value of slope, W_N has a deep minimum. This pattern is a general characteristic over a wide range of parameters, although the W_N minimum becomes quite shallow for smaller N. Thus, slopes that are either greater than at the maximum, or less than at the minimum, correspond to both lesser yield and greater risk than at the respective extrema; only slopes between these two points correspond to solutions of maximal yield.

With these insights, the slope can be eliminated and y_N obtained directly as a function of W_N, as plotted in Fig. 4.3; the Weekender example, $b_- = 0.01$, $b_+ = 0.10$, $N = 10^3$, is also shown. The latter might be typical of a weekend casino trip, with trip capital of a hundred base bets; the former of a "lifetime" of extensive play, with trip capital of a thousand base bets. Also note that the effective yield may in fact be negative for sufficiently low risks, although always greater than that without counting.

Details of the effective yield vs. risk curves vary with the choice of parameters, although the family of curves retain the general shape shown here. Displaying these variations in full detail is difficult, since the parameters form a three-dimensional family. Nevertheless, some insights can be gained from contour plots over the two-dimensional $b_\pm$ plane. First, Fig. 9.6 shows the end points, of greatest and least risk,

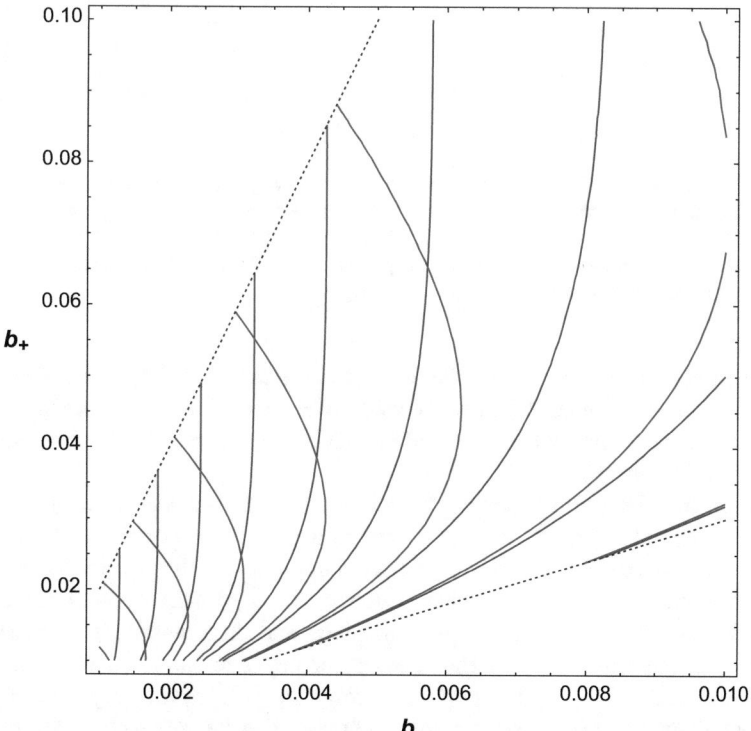

Fig. 9.6 Contours of minimum risk (blue) and maximum risk (red) vs. $b_\pm$

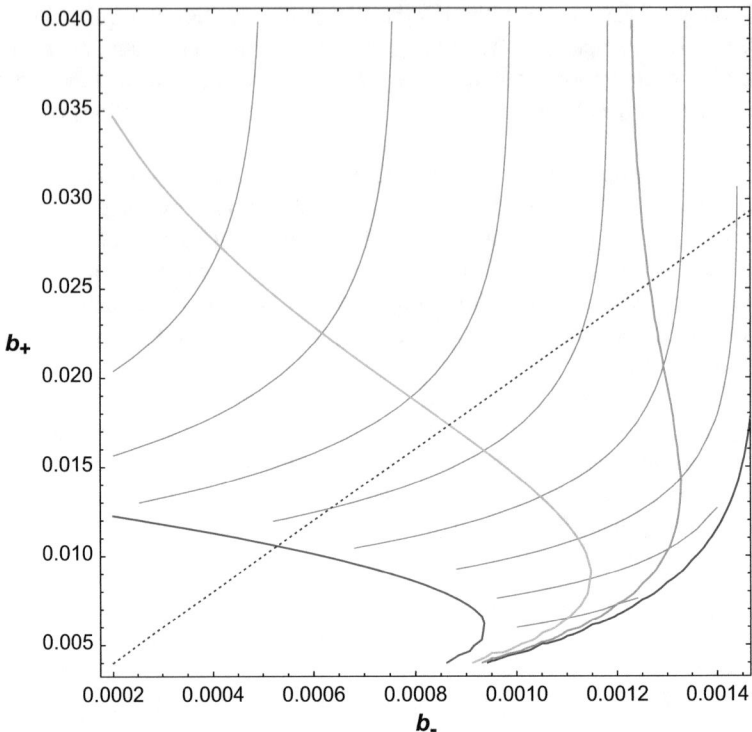

Fig. 9.7 Contours of effective yield vs. $b_\pm$, bounded between curves of minimum risk (on the right, blue) and maximum risk (on the left, red). The curve of HJY points is in green; the curve of MM points, in orange, lies closer than HJY to the minimum risk boundary and requires larger bets for an equivalent yield. Risk is fixed at the Kelly value, e^{-2}. The dotted line denotes the spread of 20, with spreads less than that to the lines's lower right

for the Lifetimer example. Then, Fig. 9.7 shows the contours of effective yield, with the risk chosen at the so-called Kelly value, $W_N = w_K \equiv e^{-2} = 0.1353$. These contours are bounded by the curves along which w_K intersects the maximum and minimum risk surfaces.

So far, the upper and lower bet bounds are regarded as chosen independently: the lower bound is held at the table minimum with the upper bound as large as practicable. Indeed, Fig. 9.7 shows for a fixed upper bound that y_N decreases as the lower bound is raised; and that for fixed lower bound y_N increases with increasing upper bound. But an emphasis on camouflage and on maintaining a predefined spread suggests instead fixing the *ratio*, β, of upper to lower bounds. Then the geometric mean of the bounds could be chosen so as to maximize y_N, even if the resulting lower bound exceeds the table minimum. The maximum of y_N under these circumstances occurs for the remarkably simple condition $s\,q = 1$, which implicitly determines s. Even more remarkably, the maximal condition is independent of the functional forms of y_N and W_N and so, in particular, is independent of N!

This condition was first emphasized, independently, by Harris, by Janecek, and by Yamashita (HJY (1997)). But HJY arrived at it by adopting quite different optimization criteria than here. Harris (1997), following Sileo (1992) and Schlesinger (2005), instead posited that a betting optimum maximizes the dimensionless ratio $\langle\langle B\ R\ \rangle\rangle^2/\langle\langle B^2\sigma^2\rangle\rangle$ at fixed spread (actually, yet equivalently, Harris (1997) minimized the inverse ratio); risk is not a determinant in his criterion. The square-root of this ratio is just the ROI (the yield per unit of typical bet size) from Sects. 4.1 and 6.1; maximal return on investment is a natural criterion for a betting optimum. Computed solutions of the HJY condition, also plotted in Fig. 9.7, are a curve through the contours of effective yield. The fact that a ray through the origin of Fig. 9.7 (a line of constant spread) is tangent to a contour at the HJY point provides a visible geometric verification that the point maximizes effective yield for that spread.

But Fig. 9.7 also shows that a continuum of optimal bet patterns, depending on risk, exists in addition to the HJY point. They range from the smallest risk and effective yield, which might be the choice of a casual, lightly capitalized player over a short session, up to the largest risk and effective yield, which could be the choice of a dedicated, well-capitalized player tolerant of occasional ruined sessions to gain greater long-term winnings. But neither end of the continuum is robust: at the lower end, a small increase in risk produces a much larger increase in effective yield; and at the upper end a small decrease in effective yield produces a much larger decrease in risk. A better balance of effective yield and risk occurs at intermediate points.

Since Player would ideally like to both minimize risk and maximize effective yield, one way to make this tradeoff is to maximize the ratio y_N/W_N, the effective yield per unit of risk. Although difficult to pursue analytically, the maximization is straightforward computationally and leads to a well-defined intermediate point, distinct from that of HJY; call it the "minmax," or "MM," point. Whereas the HJY point is at a rather high effective yield as well as risk, the MM point typically occurs at much lower values of both. Thus the former may be an appropriate choice for the dedicated player, such as the Lifetimer example of Sect. 4.1, while the latter seems more suitable for the casual player, such as the Weekender example.

9.1.7   Yield When the Bet Size Is Discrete; Wong Benchmark Betting

The optimal bet strategy derived in the previous subsection calls for a bet size that is a ramp function of the expected return. Clearly, however, bets in actual practice must be sized as a discrete multiple (an integer, or half-integer if the casino has chips of that denomination) of the minimum bet. Thus the ramp of Eqs. (9.37)–(9.38) must be approximated by a staircase, much like that represented in Fig. 4.1.

Although staircases with various numbers of steps and sizes can be used, their results were asserted earlier to be quite close to that of the corresponding ramp. To model this, select the Benchmark game and a bet spread of 10, and then replace the optimal ramp by a coarse staircase with just two intermediate steps: at 4 and at 7 base bets. Computational results of the performance of the staircase vs. the ramp, for both

Table 9.1 Performance of a coarse staircase vs. a ramp

Player	Bet shape	Steepness	Risk of ruin	Effective yield ratio
Lifetimer (HJY)	Ramp	0.94	0.1185	+0.01716
	Staircase	0.94	0.1213	+0.01713
Weekender (MM)	Ramp	2.80	0.1360	+0.00617
	Staircase	2.80	0.1404	+0.00603

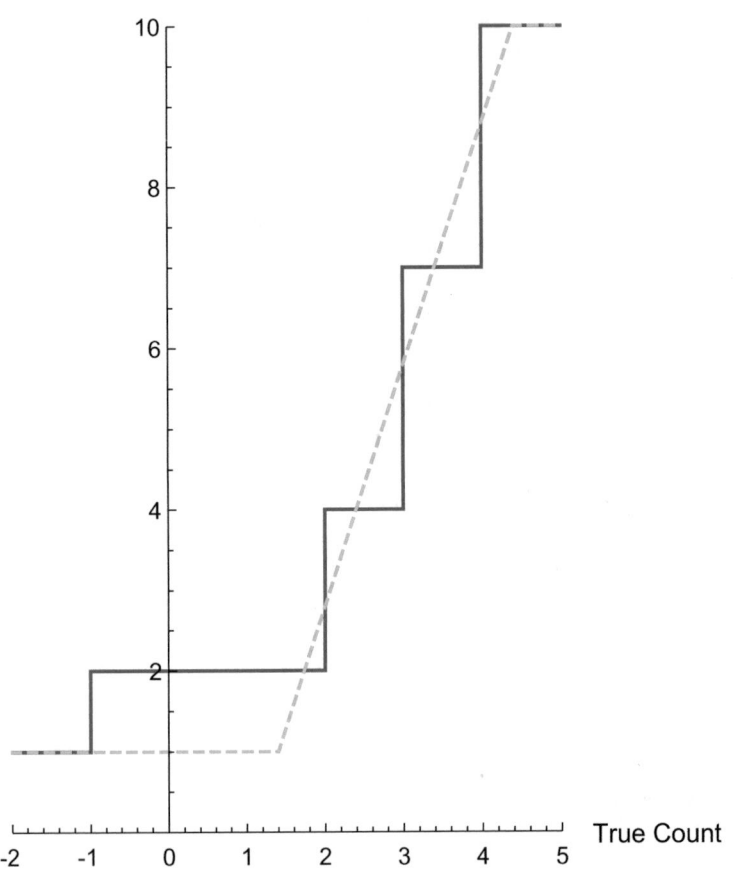

Fig. 9.8 Wong's "Benchmark" bet staircase (solid, red) and ramp modification (dashed, green)

the Lifetimer (HJY point) and the Weekender (MM point), are displayed in Table 9.1 (the ramp results are the same as in Table 4.1). Comparison demonstrates that both yield and risk are degraded by only a few percent. Other staircases, subdivided more finely, perform even more closely to the ramp.

A betting scheme recommended by Wong (1994), p. 18 and shown in Fig. 9.8 is also a staircase, but it lowers the bet another step for negative true counts and so is

Table 9.2 Performance of Wong bet function

Betting style	Yield ratio	ROI	Effective yield	Risk
Wong	0.0173	0.0041		
Wong (modified)	0.0177	0.0045		
Lifetimer			0.016	0.10
Weekender			0.014	0.37

sub-optimal. Furthermore, Wong specifies the slope of his staircase independently of both the penetration and Player's risk profile. Figure 9.8 also shows a ramp function (dashed) that gives a good representation of his staircase.

Wong applies this betting scheme in a computer simulation of his Benchmark game. He quotes a yield ratio of 0.016 and a ratio of average to minimum bet of 2.65. Wong does not exhibit a risk result, possibly because his simulation may assume infinite initial capital so that ruin is never possible. Current commercial simulation programs, however, do permit specification of an initial capital and generate ruin probabilities.

Performance measures for Wong's bet function computed in the Benchmark game are shown in Table 9.2.

The yield ratio is 0.017, with an average bet of 2.70 units, not far from the values he quotes; thus the simulation and analysis approaches give results reasonably close to each other. Also, with a typical bet of 4.2 units the return on investment is around 0.004. However, modifying Wong's function so that it bets the minimum for true counts less than +2, rather than below 0 as his does, improves both yield and ROI: a hand with negative expected return should optimally not be bet at above the minimum. In addition, Wong's yields fall sharply as the penetration is reduced, becoming negative for penetrations less than 0.5; these low penetrations need an even steeper staircase to sustain a positive yield in such unfavorable circumstances.

The effective yield and risk, with the ramp shown in Fig. 9.8, for both the Lifetimer and the Weekender, are also listed in Table 9.2. The effective yield is below that of the yield ratio because the former averages over sessions ending in ruin, whose risk is particularly high for the Weekender. The ramp is quite steep; it gives performance close to that of the HJY point—satisfactory for the Lifetimer but probably too aggressive for the Weekender.

9.2 Betting Proportional to Current Capital

Now consider bet sizes scaled to the current capital, as first suggested by Kelly (1956). Thus, set $B_N = \varphi C_N$ at the N th round and apply a criterion for choosing the scaling factor φ, independent of N; importantly, the Kelly bet size is also independent of the count! In this case, the capital changes multiplicatively rather than additively: $C_{N+1} = C_N (1 + \varphi \Omega_{N+1})$, depending on the stochastic process Ω whose values give the amount Player wins on a hand. Iterating over a sequence of rounds, the expected, or mean, value of C_N is given by

$$\langle C_N \rangle / C_0 = \left\langle \prod_{j=1}^{N} \left(1 + \varphi \Omega_j \right) \right\rangle. \tag{9.40}$$

For a simplified game where the outcome of each hand is independent of the others, such that $\langle \Omega_j \rangle = R$, this leads immediately to $\langle C_N \rangle / C_0$ increasing exponentially with N:

$$\langle C_N \rangle / C_0 = (1 + R\varphi)^N, \tag{9.41}$$

or a rate of change per round

$$\frac{1}{N} \ln \frac{\langle C_N \rangle}{C_0} = \ln(1 + R\varphi). \tag{9.42}$$

Since this rate is positive and increases monotonically with φ, it is maximized by choosing $\varphi = 1$. (When $R < 0$, the rate is negative and is minimized by $\varphi = 0$: don't bet on a losing game!) But this maximum is pathological, because it implies betting one's entire capital at every round. Only a single sequence of outcomes is successful: winning all N rounds in a row, so that the capital grows dramatically to $C_N / C_0 = 2^N$, but with the tiny probability $((1 + R)/2)^N$. Every other sequence loses, sooner or later, with $C_N = 0$; so the mean is given by Eq. (9.41). Thus the mean is not a useful measure of the capital distribution from which to select φ in any sensible way.

More promising is to examine (following Maslov and Zhang (1998), and also discussed by Ethier (2004)) the *median* of the distribution, labeled C_N^* and defined such that

$$\langle \Theta \left(\ln C_N - \ln C_N^* \right) \rangle = 1 / 2. \tag{9.43}$$

To determine C_N^*, differentiate Eq. (9.43) with respect to φ, so that

$$\frac{\partial \ln C_N^*}{\partial \varphi} = \frac{\partial}{\partial \varphi} \left\langle \sum_{j=1}^{N} \ln \left(1 + \varphi \, \Omega_j \right) \right\rangle = N \frac{\partial}{\partial \varphi} \langle \ln(1 + \varphi \, \Omega) \rangle; \tag{9.44}$$

reintegrating gives the rate of increase for the median of

$$\frac{1}{N} \ln \frac{C_N^*}{C_0} = \langle \ln(1 + \varphi \, \Omega) \rangle. \tag{9.45}$$

Thus the median capital also varies exponentially with N; but now the rate per round, the right-hand side of Eq. (9.45), has a vanishing first derivative at an intermediate value of φ. Since its second derivative there is negative definite, the value corresponds to a maximum. For $R^2 \ll 1$ the maximum is found at $\varphi \ll 1$ and so, by expanding Eq. (9.45) to second order in φ and R, is given by $\varphi_{\text{opt}} \approx R / \sigma^2 > 0$. The median capital, evaluated at that maximum, still grows exponentially,

$$\left(C_N^*/C_0\right)_{\text{opt}} \approx \exp\left(NR^2/2\sigma^2\right),\qquad(9.46)$$

but at a slower rate than when based on the mean capital.

These results also hold in the more general case where the win/loss probabilities vary from hand to hand, particularly due to a varying count. In this case, Eq. (9.45) generalizes to

$$\ln\frac{C_N^*}{C_0} = \left\langle \sum_{j=1}^{N} \ln\left(1+\varphi\,\Omega_j\right)\right\rangle,\qquad(9.47)$$

so that the maximum of C_N^* with respect to φ is approximately given by

$$\varphi_{\text{opt}} = \left\langle \sum_{j=1}^{N}\Omega_j\right\rangle / \left\langle \sum_{j=1}^{N}\Omega_j^2\right\rangle.\qquad(9.48)$$

In the absence of bet variation, the expectation of the process is just $\left\langle \sum_{j=1}^{N}\Omega_j\right\rangle = NR$, by the Invariance Theorem; and the variance is nearly independent of the true count, $\left\langle \sum_{j=1}^{N}\Omega_j^2\right\rangle \simeq N\sigma^2$.

The approach of Kelly (and of most of the subsequent literature) to selecting φ is usually described as "maximizing the expectation of the rate of capital growth." Although that terminology is loose mathematically—growth can't be ascribed to a stochastic variable, only to some measure over its distribution—the usual derivation leads to the same results as here. Our criterion, though, of maximizing the (exponential) growth rate of the *median* of the distribution seems more precise and better justified.

Note, though, that the growth rate of the median turns negative (i.e., the median shrinks) for fractions larger than about $2R$, despite the mean being positive; the rate heads towards $-\infty$ as the fraction tends towards one. The larger bets increase the risk of losing streaks eating into Player's capital; and as $\varphi \to 1$ Player is ruined very quickly with almost 100% probability. Also notice that the exponential character of the growth is discernible only for $N > \left\langle\langle\Theta\ (R)\ R^2/\sigma^2\rangle\right\rangle^{-1}$, which might be 10^4 or more; for fewer rounds, the increase still looks linear. This property suggests the mixed betting strategy discussed in the next section.

The so-called Kelly betting (there is little agreement on terminology in the literature: other labels sometimes used include scaled, proportional, multiplicative, and geometric) seems to have the remarkable advantage that Player is never ruined, in the sense of losing all his money. If he experiences a losing streak, so that his capital shrinks, he is directed merely to reduce his bet sizes accordingly and to continue play. But a bet size that varies continuously towards zero is clearly not allowed in a casino: bets can only be integer multiples of a nonzero table minimum, B_-. Rather, "ruin" should be interpreted as capital so low that the prescribed fractional bet can't be placed.

Once ruin is reconsidered in this way, analysis of optimal Kelly betting can proceed in analogy with Sect. 9.1. Instead of wins/losses being additive/subtractive to capital, here it's to the *log* of capital. And instead of ruin occurring when $C = 0$, here it's when $C = C_- \equiv B_-/\varphi$. With these analogies, a drifting diffusion equation like Eq. (9.32) also applies here, but after modification to the drift coefficients, the replacement of the C_0 variable by $\ln C_0$, and the generalized ruin boundary condition. Thus, the risk satisfies

$$\left[\frac{\partial}{\partial N} - V \frac{\partial}{\partial \ln C_0} - \frac{1}{2}\sigma^2 \frac{\partial^2}{\partial (\ln C_0)^2} \right] W_N = 0, \tag{9.49}$$

with drift $V \equiv \langle \ln(1+\varphi\,\Omega) \rangle$; the same differential equation determines the median capital. The appropriate solutions, at least in the limit of very large N, become

$$W = \exp\left(-2\frac{V}{\sigma^2} \ln \frac{C_0}{C_-} \right), \quad \frac{1}{N} \ln \frac{C_N^*}{C_0} = V\,(1 - W). \tag{9.50}$$

Again maximizing $\ln\left(C_N^*/C_0 \right)$ with respect to φ, but now subject to a fixed risk, leads to the same optimal Kelly fraction as before. The maximal rate, however, is reduced by the survival factor. With V evaluated at the maximum, the risk becomes

$$W_{\mathrm{opt}} = \exp\left[-\frac{R^2}{\sigma^4} \ln \frac{C_0\,R}{B_-\,\sigma^2} \right] \tag{9.51}$$

which, under typical conditions, is extremely small.

In blackjack, of course, $R < 0$ in the absence of counting, so Kelly betting is just as ruinous as with additive betting.

Another frequently voiced criticism of Kelly betting is that it leads to wide swings in capital, even though the median is growing at a maximal rate. To reduce the size of the fluctuations about the median, some authors (and a number of practitioners) resort to so-called "fractional Kelly:" the bet is sized at less than the optimal fraction of the current capital (i.e., "full Kelly"), perhaps half or even a quarter of the optimum.

Just as the median (or equivalently the expected logarithm) of the capital can easily be derived, so the variance of log capital is also readily obtained:

$$\mathrm{var} \equiv \left\langle \left(\ln \frac{C}{C_0} \right)^2 \right\rangle - \left\langle \ln \frac{C}{C_0} \right\rangle^2 \cong N\sigma^2 \left(\frac{1}{2} \ln \frac{1+\varphi}{1-\varphi} \right)^2, \tag{9.52}$$

which rises monotonically with φ from the origin, initially as φ^2. Thus a lower bet fraction than the optimum does reduce the variance. At the same time, a measure of return on investment, such as $\langle \ln C /C_0 \rangle/\sqrt{\mathrm{var}}$, increases slowly with decreasing φ; but does not exhibit a peak, so there is no optimal fraction of Kelly based on such a measure. Kelly bettors are forced to use their own judgment in selecting a reduced Kelly fraction.

On the other hand, criticism of full Kelly based on the size of its capital fluctuations almost always overlooks the concept of ruin in Kelly betting. Should a wide downward swing in capital occur, Player might encounter the ruin threshold and simply start over with a replenished stake, rather than continue to play with a tiny one and even tinier bets. Recovering from a situation of sharply diminished (but not ruined) capital usually takes a dismayingly long playing time, despite the exponentially increasing median. Another technique to reduce variance is to play multiple hands simultaneously, as discussed in Sect. 9.3.

9.2.1 Mixed Additive and Multiplicative Betting

Another and even more obvious shortcoming of Kelly betting, as a practical casino technique, is that it's predicated on bet sizes along a continuum, whereas of course actual bets can only be in chips of discrete denominations. As a consequence, it's not possible to rescale the bet size after every hand, since total capital (and its associated Kelly bet) has likely changed by only a tiny fraction. More reasonable would be a mixed strategy, where Player reassesses his capital only occasionally, such as at reshuffles or even just at Dealer shift changes, and rescales his bet size to the extent warranted by his current capital; between such adjustments Player bets additively, based on his capital at the start of the period.

A simple model of such a mixed betting strategy assumes an initial period of N rounds of additive betting, with capital drifting from C_0 to C_1; followed by a rescaling of bet sizes in proportion to the ratio C_1/C_0; and then by another period of N additive rounds with capital drifting from C_1 to C_2. The distribution of C_2 is obtained by convolving the two periods of drift over all intermediate values of C_1, in analogy with Eq. (9.35). Although the C_1 integration is now too complex to be obtained in closed form, it is readily computed; an example for one representative choice of parameters is shown in Fig. 9.9, along with the distribution after the same $2N$ rounds without any bet rescaling.

The distribution with Kelly rescaling, compared to that without, has an increased yield. Despite its greater risk, from the increased weight at low C_2, the considerably greater weight at higher C_2 leads to an overall higher mean.

Nonetheless, the model still overlooks the reality of bets needing to be discrete and larger than a minimum.

9.3 Betting When Playing Multiple Simultaneous Hands

We've already covered, in Sect. 7.1.4, the expectation and variance of the return when playing H hands simultaneously. Here we examine the expectation and variance of the change per round of total capital, beginning with the broad assumption that each of the hands, labeled h, takes a distinct bet B_h. Thus

$$\langle \delta C \rangle = \left\langle \sum_{h=1}^{H} B_h \Omega^h \right\rangle = \left(\sum_{h=1}^{H} B_h \right) R \equiv \hat{B} R,$$

$$\left\langle (\delta C)^2 \right\rangle - \langle \delta C \rangle^2 = \left\langle \left(\sum_{h=1}^{H} B_h \Omega^h \right)^2 \right\rangle - \left\langle \sum_{h=1}^{H} B_h \Omega^h \right\rangle^2 \tag{9.53}$$

$$= \left(\sum_{h=1}^{H} B_h^2 \right) \sigma^2 + \left[\left(\sum_{h=1}^{H} B_h \right)^2 - \sum_{h=1}^{H} B_h^2 \right] \Gamma,$$

where the stochastic process Ω was introduced in Sect. 9.2, and $\hat{B}$ is the total amount bet on the round. These results are reminiscent of those in Sect. 7.1.4.

Now consider the bet optimization criterion of Sect. 9.1, that of maximizing the yield with respect to each of the H bets separately, while holding the risk of ruin constant. But, following an argument similar to that of Sect. 9.1.5, the yield and risk of ruin depend on the B_h only via the first two moments of the change in capital per round, as per Eq. (9.53). So the functional maximization must lead to a set of H simultaneous linear equations, in the variables

$$\delta \langle \delta\, C \rangle / \delta\, B_h = R, \quad \delta \left(\left\langle (\delta C)^2 \right\rangle - \langle \delta C \rangle^2 \right) \Big/ \delta B_h = 2 B_h \sigma^2 + 2(\hat{B} - B_h) \Gamma. \tag{9.54}$$

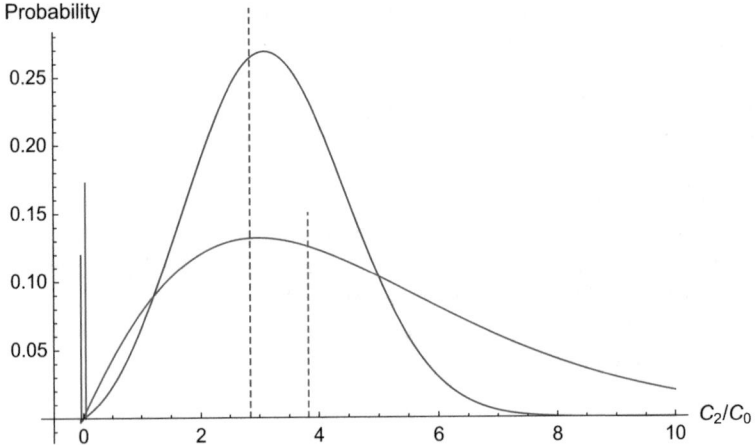

Fig. 9.9 Distribution of capital, C_2, in units of C_0 and with a representative choice of game and betting parameters, after $2\,N$ rounds: blue curve, no bet rescaling; red curve, Kelly rescaling after N rounds. The dashed vertical bars mark the means of the distribution of corresponding color; the mean for the mixed strategy (red) shows a considerable improvement over that with no rescaling (blue). The solid vertical bars at the origin display the risk of ruin for the distribution of corresponding color; Kelly rescaling moderately increases the risk

Since each B_h is only linked to $\hat{B}$ rather than to any of the other $B_{h' \neq h}$, the equations separate and lead to B_h independent of h. Solving the maximization equation, then, for the bet size per hand gives $B = \hat{B}/H \propto HR/\sigma_H^2$, where σ_H^2 is defined in Eq. (7.20). The derivation shows that optimal betting with multiple hands requires the same bet size for each; that the optimal size is again directly proportional to the expected return on the round; and that the factor of proportionality now contains the covariance between different hands added to the variance of a single hand.

The above analysis also gives a simple result for the yield and risk of ruin with multiple hands. These still satisfy a diffusion equation like Eq. (9.32), but now with the coefficients modified to reflect Eq. (9.53):

$$\langle\langle BR \rangle\rangle \longrightarrow H\langle\langle BR \rangle\rangle = \left\langle\left\langle \hat{B}R \right\rangle\right\rangle,$$

$$\langle\langle B^2\sigma^2 \rangle\rangle \longrightarrow \langle\langle B^2\sigma_H^2 \rangle\rangle = \left\langle\left\langle \hat{B}^2(\sigma^2 + (H-1))\Gamma \right\rangle\right\rangle \Big/ H. \tag{9.55}$$

Although the distribution of total capital drifts at a rate independent of H, it broadens at a rate that decreases with H only slightly slower than $H^{-1/2}$. Thus, spreading a given total bet evenly among multiple seats is a useful way to lower the variance of the outcome, and hence to lower the risk of ruin.

The Kelly betting method is similarly improved through multiple hands, but also allows the advantageous technique of capital redistribution. In this procedure, Player initially splits his total capital evenly among his seats, creating sub-pools, and at each seat bets a uniform fraction φ_H of the sub-pool capital. After this and every subsequent round, Player then re-pools his capital, again divides it evenly (or as evenly as feasible) among the seats, and repeats the betting process. In effect, the scheme uses the winning seats to replenish the losing ones. A scheme of this kind has been proposed for financial investing by Maslov and Zhang (1998).

Designate the capital of the h th seat at the end of the n th round as C_n^h, so that the redistributed capital per seat is

$$C_n = 1/H \sum_{h=1}^{H} C_n^h \tag{9.56}$$

Then, at the h th seat the capital following the next round becomes

$$C_{n+1}^h = \left(1 + \varphi_H \, \Omega_{n+1}^h\right) \bar{C}_n. \tag{9.57}$$

The stochastic process Ω_n^h is uncorrelated from other rounds but, as seen in Sect. 7.1.4, has a non-vanishing covariance with the other hands on the same round. Thus, after N rounds the redistributed capital per seat has evolved to

$$\bar{C}_N / \bar{C}_0 = \prod_{n=1}^{N} \left(1 + \varphi_H \, \bar{\Omega}_n \right), \tag{9.58}$$

where $\bar{\Omega}_n \equiv H^{-1} \sum_{h=1}^{H} \Omega_n^h$. This is again a multiplicative process, like that for a single seat, and the mean of the redistributed capital is

$$\langle \bar{C}_N \rangle = \bar{C}_0 \, (1 + R \, \varphi_H)^N . \tag{9.59}$$

On the other hand, the median $\bar{C}_N^*$ satisfies

$$\frac{1}{N} \ln \frac{\bar{C}_N^*}{\bar{C}_0} = \frac{1}{N} \sum_{n=1}^{N} \langle \ln \left(1 + \varphi_H \, \bar{\Omega}_n \right) \rangle . \tag{9.60}$$

Anticipating that the median rate's maximum occurs for $\varphi_H \ll 1$ and $R \ll 1$, expand in this limit:

$$\frac{1}{N} \ln \frac{\bar{C}_N^*}{\bar{C}_0} = \frac{1}{N} \sum_{n=1}^{N} \left(\varphi_H \langle \bar{\Omega}_n \rangle - \frac{1}{2} \varphi_H^2 \langle \bar{\Omega}_n^2 \rangle + \cdots \right).$$

$$= \varphi_H \, R - \frac{1}{2} \varphi_H^2 \left[\sigma^2 + (H - 1) \Gamma \right] \Big/ H + \cdots \tag{9.61}$$

Thus the maximum rate now occurs for the proportionality fraction $\varphi_H = H^2 R \big/ \sigma_H^2$, where the effect of the higher order terms in Eq. (9.61) can be shown to be negligible. At this fraction,

$$\frac{1}{N} \ln \frac{\bar{C}_N^*}{\bar{C}_0} = \frac{H^2 R^2}{2\sigma_H^2} . \tag{9.62}$$

Remarkably, both the maximal rate of growth of the median capital, and the betting fraction that achieves it, have increased as a result of playing H hands simultaneously and redistributing capital among them, as compared with Kelly betting with the total capital at a single seat. As seen in Eq. (9.61), the single-hand variance is reduced by a factor of H, which in turn boosts the growth rate, although the effect is partially offset by the covariance. Thus, in the context of Kelly betting, playing multiple hands and systematically redistributing capital improves performance versus that of just a single hand with the same total capital.

The variance of log capital in this scheme, for small but otherwise arbitrary values of φ_H, is var $= N \left[\sigma^2 + (H - 1) \, \Gamma \right] \varphi_H^2 / H$; it too would increase, proportionally, if the optimal bet fraction of the previous paragraph were employed. But since typical numerical values are $\Gamma \approx 0.47 < \sigma^2 \approx 1.26$, the coefficient of φ_H^2 in the variance expression decreases with increasing H. As a result, if Player maintains the same bet fraction as if playing only a single hand, namely $\varphi_1 = R / \sigma^2$, rather than the larger value from the optimality criterion, then playing several hands would both reduce the variance as well as raise the median's rate of increase! In this technique, the median capital and the variance become

$$\ln\left(\frac{\bar{C}_N^*}{C_0}\right) = \frac{N R^2}{2\sigma^2}\left[1 + \frac{H-1}{H}\left(1 - \frac{\Gamma}{\sigma^2}\right)\right], \quad \text{var} = \frac{N R^2}{\sigma^2}\left[1 - \frac{H-1}{H}\left(1 - \frac{\Gamma}{\sigma^2}\right)\right].$$

Even with only two seats, the median rate is raised by more than 30% while the variance is reduced by the same percentage; with still more simultaneous hands the factors can exceed 50%.

9.4   Back-Counting and Table-Hopping

Section 4.3 has described the playing technique of table-hopping and its variant, back-counting. These maneuvers offer a significant improvement in yield, quantified in Table 4.5. Here the analysis underlying the Table (following Werthamer (2006) and (2008)) is described for entry, exit, and departure, individually and in combination.

9.4.1   Entry

Begin with certain assumptions about Player and the table he's at:

- the game is played with more than one deck, $D > 1$, so that the expected return R_0 for the first round after a shuffle is negative and has variance σ^2;
- the cards are reshuffled after a shoe penetration F;
- Player is tracking the dealt cards using one of the usual balanced counting methods, such that the true count immediately after a shuffle is $\gamma = 0$ and the expected return becomes positive, $R > 0$, for true counts greater than a positive cross-over, $\gamma \geq \gamma_0 > 0$;
- entry is allowed between any successive rounds, not just at shuffles.

The back-counter enters the game only when the true count first reaches the threshold $\gamma_E \geq \gamma_0$. Thus we need the probability of this occurring, conditional on the true count being zero immediately following a shuffle. But this probability is mathematically congruent to the probability of ruin, i.e., of Player's capital C first reaching zero from its initial value of C_0. Equation (9.6) has shown that the probability of ruin first occurring at round N, with a bet B per round, is

$$L_N = \left(C_0/N B\right)\rho_N(0), \tag{9.63}$$

where (Eq. (9.19))

$$\rho_N(C) \equiv \frac{1}{(2\pi N B^2\sigma^2)^{1/2}}\exp\left[-\frac{(C - C_0 - N B R)^2}{2N B^2\sigma^2}\right] \tag{9.64}$$

is the distribution of capital after N rounds, in the absence of ruin. Furthermore, L_N can be shown to satisfy the drifting diffusion equation (like Eq. (9.32))

$$\left(\frac{\partial}{\partial N} - BR \frac{\partial}{\partial C_0} - \frac{1}{2} B^2 \sigma^2 \frac{\partial^2}{\partial C_0^2} \right) L_N = 0. \tag{9.65}$$

In parallel, the distribution of true counts was shown (Eqs. (8.24) and (8.7), with $\Lambda = 1$) to be

$$\rho(\gamma, \tau) = \frac{1}{\sqrt{2 \pi \tau}} \exp \left(-\frac{\gamma^2}{2\tau} \right), \tag{9.66}$$

contingent on the initial condition of zero true count at zero penetration. As a modification from the notation of Sect. 8.1.1, the parameter $\tau \equiv 52 f / D \, (1 - f) = (52\Delta)^2$ is used; the extra factor of 52 appears because the true count is defined as running count *per deck* un-dealt. We call τ "time" because the distribution can be verified to satisfy the diffusion equation

$$\left(\frac{\partial}{\partial \tau} - \frac{1}{2} \frac{\partial^2}{\partial \gamma^2} \right) \rho(\gamma, \tau) = 0, \tag{9.67}$$

where τ plays the mathematical role that time does in the physical diffusion process. The fact that τ is a non-linear function of depth, which of course *does* increase proportionally to real time, reflects the character of the underlying random card-dealing process: the true count, which is always zero just after a shuffle and then random walks, must return to zero if/when all cards in the shoe are dealt. Such a constrained process is sometimes called a Brownian bridge.

Thus there is a close analogy in their evolution between the distributions of capital and of true count (compare Eqs. (9.65) and (9.67)), which gives the probability of the true count first reaching γ_E at "time" τ_E as

$$\rho_1(\gamma_E, \tau_E) = (\gamma_E / \tau_E) \, \rho(\gamma_E, \tau_E). \tag{9.68}$$

Furthermore, the distribution $\rho_E(\gamma, \tau)$ of true count γ at any "time" τ prior to entry must still satisfy Eq. (9.67) but now with the boundary condition $\rho_E(\gamma_E, \tau) = 0$. This boundary condition is the analog of that for "survival," from Sect. (9.1); it is satisfied by including in the solution a "reflection" about γ_E:

$$\rho_E(\gamma, \tau) = (\rho(\gamma, \tau) - \rho(2\gamma_E - \gamma, \tau)) \, \Theta(\gamma_E - \gamma), \tag{9.69}$$

where Θ is again the unit step function. Note that any solution to the diffusion equation (and many other differential equations) is unique if it also satisfies the initial/boundary conditions.

This pair of expressions is further justified by showing that they satisfy conservation of probability. Thus the total probability of *having* entered at any τ_E prior to τ is, from Eq. (9.68),

$$\tilde{P}_E(\tau) = \int_0^\tau d\tau_E \, \rho_1(\gamma_E, \tau_E) = 2\mathrm{erf}\left(\frac{\gamma_E}{\sqrt{\tau}}\right), \tag{9.70}$$

whereas the total probability of *not* having entered prior to τ is, from Eq. (9.69),

$$P_E(\tau) = \int_{-\infty}^{\infty} d\gamma \, \rho_E(\gamma, \tau) = 1 - 2\mathrm{erf}\left(\frac{\gamma_E}{\sqrt{\tau}}\right). \tag{9.71}$$

These indeed sum to unity, as required by conservation of probability.

In addition, the two expressions are also linked by the iterative "chain rule,"

$$\frac{1}{\tau_E}\int_0^{\tau_E} d\tau \int_{-\infty}^{\infty} d\gamma \, \rho_1(\gamma_E - \gamma, \tau_E - \tau) \, \rho_E(\gamma, \tau) = \rho_1(\gamma_E, \tau_E); \tag{9.72}$$

the integration steps needed for its proof also serve as a prototype for similar convolutions in succeeding sections, so are detailed in Appendix 2.

Next, given that the true count γ_E is first reached at τ_E, the subsequent evolution of the true count distribution is again a solution of Eq. (9.67) with those initial conditions, specifically $\rho(\gamma - \gamma_E, \tau - \tau_E)$, a generalization of Eq. (9.66). Then the distribution of true counts experienced by a back-counter who bets only *after* the true count has first reached γ_E is the convolution with the probability of Eq. (9.68),

$$\int_0^\tau d\tau_E \, \rho(\gamma - \gamma_E, \tau - \tau_E) \, \rho_1(\gamma_E, \tau_E) = \rho(\Gamma_E, \tau), \tag{9.73}$$

$$\Gamma_E \equiv |\gamma - \gamma_E| + \gamma_E$$

carrying out the integration uses the same methods as in Appendix 2. The back-counter, once he's entered, then bets an amount $B(\gamma)$ based on the true count, and plays until the next reshuffle at penetration F. With the true count distribution of Eq. (9.73), his yield is

$$Y_E = \int_{-\infty}^{\infty} d\gamma \, B(\gamma) \, R(\gamma) \, P(\gamma), \qquad P(\gamma) \equiv \frac{1}{F}\int_0^F df \, \rho(\Gamma_E, \tau). \tag{9.74}$$

The entry threshold γ_E is chosen so as to maximize the yield. The distribution $P(\gamma)$ is graphed in Fig. 9.10.

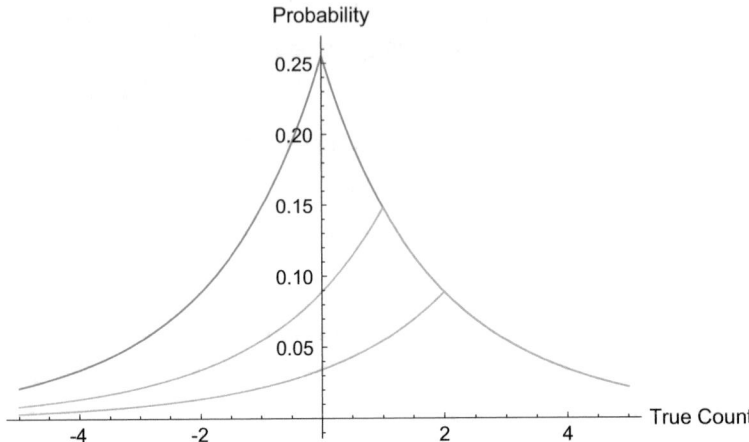

Fig. 9.10 Distributions of true counts at entry and subsequently: entry at shuffle (blue curve), entry at true counts +1 (orange), and +2 (green); 6 decks. Notice how the likelihood of true counts <1, with negative return, is dramatically reduced via the entry maneuver. To interactively select other values for these parameters, browse to the Wolfram Cloud website www.wolfr. am/BlackjackScience and select the appropriate value. This figure is copyright by N. Richard Werthamer and used by permission

9.4.2 Entry and Exit

Thus far we've assumed that the back-counter, once he's entered the game, remains in until the shoe is reshuffled. But some authorities (e.g., Vancura and Fuchs (2016) p. 132) suggest that he leave the table—we'll call it exit—when the true count drops below the roughly +1 value at which the expected return changes from positive to negative. We'll analyze the more general situation of exit at a true count γ_X, and choose it jointly with γ_E to maximize the yield.

Without exit, the distribution of true counts following entry is given by Eq. (9.68), and satisfies the diffusion equation, Eq. (9.67). With exit, the modified distribution $\rho_{EX}(\gamma, \tau)$ for $\gamma \geq \gamma_X$ at time τ must still satisfy Eq. (9.67) but now with the boundary condition $\rho_{EX}(\gamma_X, \tau) = 0$. The solution, analogous to Eq. (9.69), is accomplished with a reflection about γ_X:

$$\rho_{EX}(\gamma, \tau) = [\rho(\Gamma_E, \tau) - \rho(\mid 2\gamma_X - \gamma - \gamma_E \mid + \gamma_E, \tau)]\, \Theta(\gamma - \gamma_X).$$
(9.75)

Equations (9.75) and (9.69) are compared schematically in Fig. 9.11, for the same game parameters as in Fig. 9.10.

Furthermore, the probability of exiting at "time" τ_X, conditional on entry at τ_E, is the convolution of their respective probabilities,

$$P_X(\tau_X) = \int_0^{\tau_X} d\tau_E \, \rho_1(\gamma_X - \gamma_E, \tau_X - \tau_E) \, \rho_1(\gamma_E, \tau_E)$$

$$= \rho_1(2\gamma_E - \gamma_X, \tau_X). \tag{9.76}$$

These two expressions, like Eqs. (9.70) and (9.71) above, similarly satisfy conservation of probability but now conditional on entry:

$$\int_\tau^{\tau_E} d\tau_X \, P_X(\tau_X) + \int_{-\infty}^{\infty} d\gamma \, \rho_{EX}(\gamma, \tau) = 2\mathrm{erf}\left(\frac{\gamma_E}{\sqrt{\tau}}\right), \tag{9.77}$$

where the right-hand side, by Eq. (9.70), is the probability of entry prior to τ.

The yield from this table is the analog of Eq. (9.74):

$$Y_{EX}^{(1)} = \frac{1}{F} \int_0^F df \int_{-\infty}^{\infty} d\gamma \, B(\gamma) \, R(\gamma) \, \rho_{EX}(\gamma, \tau) \tag{9.78}$$

But upon exit, a back-counter can "table-hop" to a second table with a freshly shuffled shoe and repeat the entry process; although finding an appropriate second table may realistically take some time, we here assume the switch to be instantaneous.

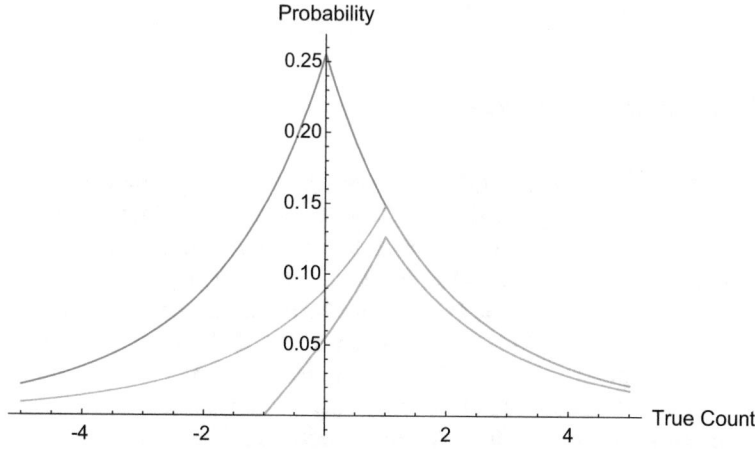

Fig. 9.11   Distribution of true counts, with entry and exit. The blue curve is without any entry or exit; the orange curve is with just entry at a count of +1; and green is for both entry (at +1) and exit (at −1), all for 6 decks. Notice how the addition of exit removes still more of the weight for negative returns vs. entry alone. To interactively select other values for these parameters, browse to the Wolfram Cloud website www.wolfr.am/BlackjackScience and select the appropriate value. This figure is copyright by N. Richard Werthamer and used by permission

Hence the most appropriate assessment of his yield is to include the cash flow from the second table during those rounds of the first that follow the exit and precede its reshuffle. Thus the yield from the second table alone, weighting with the probability of exit from the first, is

$$
Y_{EX}^{(2)} = \int_0^{\tau_F} d\tau_X \, P_X (\tau_X) \, \frac{1}{F} \int_0^{F-f_X} df \int_{-\infty}^{\infty} d\gamma \; B(\gamma) \, R(\gamma) \, \rho_{EX}(\gamma, \tau). \qquad (9.79)
$$

By combining the yields from the two tables, and carrying out the intermediate τ_X integration, the total yield becomes

$$
Y_{EX} = Y_{EX}^{(1)} + Y_{EX}^{(2)}
$$

$$
= \frac{1}{F} \int_0^F df \left[1 + 2\mathrm{erf} \left(\frac{2\gamma_E - \gamma_X}{\sqrt{\tilde{\tau}}} \right) \right] \int_{-\infty}^{\infty} d\gamma \; B(\gamma) \, R(\gamma) \, \rho_{EX}(\gamma, \tau),
$$

$$
(9.80)
$$

where $\tilde{\tau} \equiv 52 \, (F-f)/D \, (1-F+f)$. Although contributions from additional tables beyond the second should in principle be included, in practice these are negligible: the probability of exit from the second table is quite small and the number of rounds played at a third is typically too few to generate much additional value. The optimal entry and exit thresholds are determined, as in the entry-only case, by jointly maximizing the total yield.

9.4.3 Entry and Departure

Separately from a possible exit *following* entry, the table-hopper may independently choose to leave the table *prior* to entry. This might occur, for example, if much of the shoe has been dealt without reaching the entry threshold; or if the true count becomes decidedly negative and the probability of it swinging sufficiently positive to trigger entry is correspondingly low. Ideally, the departure decision should be based on a combination of these two circumstances. But Player decision-making predicated on both true count and depth parameters together seems difficult to master; rather, we consider departure based only on true count.

Exit, occurring after entry by definition, imposes the single boundary constraint $\gamma \geq \gamma_X$, satisfied as per Eq. (9.75) by a reflection about γ_X. Departure, occurring alternatively to entry, instead imposes the two simultaneous boundary constraints $\gamma_D \leq \gamma \leq \gamma_E$, which requires the more complex structure of an infinite array of reflections. A real-world example is that if one looks in a mirror one sees a single reflection; but if one instead looks in a mirror with a second one parallel to and facing it, one sees infinitely many, receding reflections.

Thus the solution of the diffusion equation subject to the two boundary constraints is the generalization of Eq. (9.69),

$$\rho_{ED}(\gamma, \tau) = \sum_{m=-\infty}^{\infty} [\rho(\gamma - a_m) - \rho(\gamma - 2\gamma_E - a_m)],$$

(9.81)

$$a_m \equiv 2m(\gamma_E - \gamma_D).$$

Then the probability of entry at τ_E is given by the generalization of the chain rule, Eq. (9.73),

$$\rho_{1D}(\gamma_E, \tau_E) = \frac{1}{\tau_E} \int_0^{\tau_E} d\tau \int_{\gamma_D}^{\gamma_E} d\gamma \, \rho_1(\gamma_E - \gamma, \tau_E - \tau) \, \rho_{ED}(\gamma, \tau)$$

$$= \frac{\gamma_E}{\tau_E} \rho(\gamma_E, \tau_E) - \frac{\gamma_E - \gamma_D}{\tau_E} \sum_{m=1}^{\infty} [\rho(a_m - \gamma_E, \tau_E)$$

(9.82)

$$- \rho(a_m + \gamma_E, \tau_E)];$$

the integrations use the same methods as for the chain rule, although the algebra is lengthier. Since the terms in the sum decrease exponentially with increasing m, a reasonable approximation is to keep only the larger of the two $m = 1$ terms; the result is

$$\rho_{1D}(\gamma_E, \tau_E) \cong \frac{\gamma_E}{\tau_E} \rho(\gamma_E, \tau_E) - \frac{\gamma_E - \gamma_D}{\tau_E} \rho(2\gamma_D - \gamma_E, \tau_E).$$

(9.83)

The corresponding result for the distribution of γ at τ subsequent to entry, generalizing Eq. (9.73), is

$$\rho_{ED}(\gamma, \tau) = \int_0^{\tau} d\tau_E \, \rho(\gamma - \gamma_E, \tau - \tau_E) \, \rho_{1D}(\gamma_E, \tau_E)$$

$$= \rho(\Gamma_E, \tau) - \sum_{m=1}^{\infty} \frac{\gamma_E - \gamma_D}{a_m - \gamma_E} [\rho(|\gamma - \gamma_E| + a_m - \gamma_E, \tau)$$

(9.84)

$$- \rho(|\gamma + \gamma_E| + a_m + \gamma_E, \tau)]$$

$$\cong \rho(\Gamma_E, \tau) - \frac{\gamma_E - \gamma_D}{\gamma_E - 2\gamma_D} \rho(\Gamma_E - 2\gamma_D, \tau).$$

Then the yield from the first table is

$$Y_{ED}^{(1)} = \frac{1}{F} \int_0^F df \int_{-\infty}^{\infty} d\gamma \, B(\gamma) \, R(\gamma) \, \rho_{ED}(\gamma, \tau). \tag{9.85}$$

Additionally, the probability of departure is the reverse of Eqs. (9.82) and (9.83),

$$\begin{aligned}
P_D(\tau_D) &= \frac{-\gamma_D}{\tau_D} \rho(\gamma_D, \tau_D) - \frac{\gamma_E - \gamma_D}{\tau_D} \sum_{m=1}^{\infty} [\, \rho(a_m + \gamma_D, \tau_D) \\
&\quad - \rho(a_m - \gamma_D, \tau_D)\,] \; P_D(\tau_D) \\
&\cong \frac{-\gamma_D}{\tau_D} \rho(\gamma_D, \tau_D) - \frac{\gamma_E - \gamma_D}{\tau_D} \rho(2\gamma_E - \gamma_D, \tau_D).
\end{aligned} \tag{9.86}$$

Then the yield from a second table following departure becomes, like Eq. (9.79),

$$Y_{ED}^{(2)} = \int_0^{\tau_F} d\tau_D \, P_D(\tau_D) \, \frac{1}{F} \int_0^{F-f_D} df \int_{-\infty}^{\infty} d\gamma \, B(\gamma) \, R(\gamma) \, \rho_{ED}(\gamma, \tau), \tag{9.87}$$

and like Eq. (9.80) the total yield from both tables is

$$\begin{aligned}
Y_{ED} &= Y_{ED}^{(1)} + Y_{ED}^{(2)} \\
&\cong \frac{1}{F} \int_0^F df \left[1 + 2\mathrm{erf}\left(\frac{-\gamma_D}{\sqrt{\tau}}\right) - 2\frac{\gamma_E - \gamma_D}{2\gamma_E - \gamma_D}\mathrm{erf}\left(\frac{2\gamma_E - \gamma_D}{\sqrt{\tau}}\right) \right] \\
&\quad \times \int_{-\infty}^{\infty} d\gamma \, B(\gamma) R(\gamma) \, \rho_{ED}(\gamma, \tau).
\end{aligned} \tag{9.88}$$

As with exit, the optimal entry and departure thresholds are determined by joint maximization of the yield.

9.4.4 Entry, Exit, and Departure

When the table-hopper employs both departure (prior to entry) and exit (subsequent to entry), the probability distribution of true counts at τ combines Eqs. (9.75) and (9.84),

$$
\rho_{EDX}(\gamma, \tau) = \int_0^{\tau} d\tau_E [\rho (\gamma - \gamma_E, \tau - \tau_E)
$$
$$
- \rho (2\gamma_X - \gamma - \gamma_E, \tau - \tau_E)] \, \rho_{1D}(\gamma_E, \tau_E)
$$
$$
= [\rho_{ED}(\gamma, \tau) - \rho_{ED}(2\gamma_X - \gamma, \tau)] \Theta (\gamma - \gamma_X), \tag{9.89}
$$

and the yield from the first table is

$$
Y_{EDX}^{(1)} = \frac{1}{F} \int_0^F df \int_{-\infty}^{\infty} d\gamma \; B(\gamma) \, R(\gamma) \, \rho_{EDX}(\gamma, \tau). \tag{9.90}
$$

The probability of departure remains independent of any possibility of later exit, so the departure contribution to the second table's yield is still given by Eq. (9.85); but the probability of exit is influenced by the possibility of previous departure, so that Eq. (9.76) generalizes to

$$
P_{XD}(\tau_X) = \int_0^{\tau_X} d\tau_E \, \rho_1 (\gamma_X - \gamma_E, \tau_X - \tau_E) \, \rho_{1D}(\gamma_E, \tau_E)
$$
$$
\cong \rho_1 (2\gamma_E - \gamma_X, \tau_X)
$$
$$
- \frac{\gamma_E - \gamma_D}{2\gamma_E - \gamma_D} \rho_1 (2\gamma_E - \gamma_X - 2\gamma_D, \tau_X). \tag{9.91}
$$

Thus the total yield from the first two tables combined, generalizing Eqs. (9.80) and (9.88), is

$$
Y_{EDX} = \frac{1}{F} \int_0^F df \left\{ 1 + 2 \left[\mathrm{erf} \left(\frac{-\gamma_D}{\sqrt{\tau}} \right) + \mathrm{erf} \left(\frac{2\gamma_E - \gamma_X}{\sqrt{\tau}} \right) \right] \right.
$$
$$
\left. - 2 \frac{\gamma_E - \gamma_D}{2\gamma_E - \gamma_D} \left[\mathrm{erf} \left(\frac{2\gamma_E - \gamma_D}{\sqrt{\tau}} \right) + \mathrm{erf} \left(\frac{2\gamma_E - \gamma_X - 2\gamma_D}{\sqrt{\tau}} \right) \right] \right\} \tag{9.92}
$$
$$
\times \int_{-\infty}^{\infty} d\gamma \; B(\gamma) \, R(\gamma) \, \rho_{EDX}(\gamma, \tau).
$$

As before, the three threshold true counts are determined jointly by maximizing the yield. Computations with Eq. (9.92), despite its apparent complexity, are feasible and examples of the resulting thresholds are listed in Table 4.5.

Appendix 1: Distribution of Player's Capital, Asymptotically for Large N

Begin with Eq. (9.12) for the Fourier transform $\hat{F}_N(\phi)$, and carry out its inverse Fourier transform to obtain an expression for $\hat{\rho}_N(C)$:

$$\hat{\rho}_N(C) = \int_{-\pi}^{\pi} \frac{1}{2\pi} \left\{ \exp\left[-i\,(c - c_0)\,\phi + N \ln v(\phi)\right] \right.$$

$$\left. - \sum_{\mu=c_0}^{N} \exp[-ic\phi + (N - \mu)\ln v(\phi)] \right\} d\phi \qquad (9.93)$$

The asymptotic limit for large N forces $\ln v(\phi)$ towards the neighborhood of its minimum value, occurring at $\phi = 0$; in that region,
$\ln v(\phi) \approx i R\phi - \sigma^2 \phi^2 / 2$; hence

$$\hat{\rho}_N(C) \approx \frac{1}{(2\pi N\sigma^2)^{1/2}} \exp\left(\frac{(c - c_0 - NR)^2}{2N\sigma^2}\right)$$

$$- \sum_{\mu=c_0}^{N} \frac{1}{(2\pi(N-\mu)\sigma^2)^{1/2}} \exp\left(\frac{(c - (N-\mu)R)^2}{2(N-\mu)\sigma^2}\right) L_\mu. \quad (9.94)$$

Next, pursue the asymptotic limit of L_μ: starting with Eq. (9.6), substitute its sequential predecessors Eqs. (9.4), (9.2), and (9.1). Then use the two Kronecker deltas to eliminate all but one of the three n_ω summations; the most convenient variable to remain is $\lambda \equiv n_+ + n_-$. Next, proceed to the asymptotic limit as in Appendix 1 of Chap. 8: replace all factorials by their asymptotic limit (the Stirling approximation again, $n! \approx \sqrt{2\pi n}\,(n/e)^n$) and change the remaining sum over the discrete variable λ into an integration over a continuous one. Then

$$L_\mu \approx \int_{c_0}^{\mu} \frac{1}{2\pi} \left(\frac{c_0^2}{(\lambda^2 - c_0^2)\,\mu(\mu - \lambda)}\right)^{1/2} \exp(E(\mu, \lambda))d\lambda, \qquad (9.95)$$

where $E(\mu, \lambda) \equiv \sum_{\pm} \left(\frac{\lambda \mp c_0}{2} \ln \frac{2\mu\xi(\pm)}{\lambda \mp c_0}\right) + (\mu - \lambda) \ln \frac{\mu\xi(0)}{\mu - \lambda}$.

Substitute back into Eq. (9.94), interchange the order of integrations, and apply the Method of Stationary Phase to the integration over μ. The stationary point of $E(\mu, \lambda)$ with respect to μ is at $\mu = \lambda/(1 - \xi(0)) \equiv \mu_0$, and in its vicinity $E(\mu, \lambda) \approx E(\mu_0, \lambda) - \lambda (\mu - \mu_0)^2 / 2\mu_0(\mu_0 - \lambda)$. With limits extended to $\pm\infty$, the μ integral becomes a Gaussian, so that

$$\hat{\rho}_N(C) \approx \frac{1}{\sqrt{2\pi\hat{N}}} \exp\left(\frac{\left(c - c_0 - \hat{N}\hat{R}\right)^2}{2\hat{N}}\right)$$

$$-\int_{c_0}^{\hat{N}} \left(\frac{c_0^2}{2\pi\lambda\left(\lambda^2 - c_0^2\right)}\right)^{1/2} \frac{1}{\sqrt{2\pi\left(\hat{N} - \lambda\right)}}$$

$$\times \exp\left(\frac{\left(c - \left(\hat{N} - \lambda\right)\hat{R}\right)^2}{2\left(\hat{N} - \lambda\right)}\right) \exp\left(E\left(\mu_0, \lambda\right)\right) d\lambda, \qquad (9.96)$$

using the more compact notation $\hat{N} \equiv N\sigma^2$, $\hat{R} \equiv R/\sigma^2$. The second exponent inside the λ integral can be rearranged as $E(\mu_0, \lambda) = -c_0\Theta(R) \ln \frac{1+\hat{R}}{1-\hat{R}} + \frac{\lambda}{2} \ln \frac{\lambda^2\left(1-\hat{R}^2\right)}{\lambda^2 - c_0^2} - \frac{c_0}{2} \ln\left(\frac{\lambda + c_0}{\lambda - c_0} \frac{1-\hat{R}}{1+\hat{R}}\right)$, which for $R^2 \ll 1$ and $\lambda \gg c_0$ approximates to $E(\mu_0, \lambda) \simeq -2c_0\hat{R}\Theta(R) - \frac{\hat{R}^2}{2\lambda}\left(\lambda - \frac{c_0}{\hat{R}}\right)^2$. Substituting these approximations back into Eq. (9.96), the exponentials can be combined and rearranged so that

$$\hat{\rho}_N(C) \approx \frac{1}{\sqrt{2\pi\hat{N}}} \exp\left(-\frac{\left(c - c_0 - \hat{N}\hat{R}\right)^2}{2\hat{N}}\right)$$

$$-\frac{1}{\sqrt{2\pi\hat{N}}} \exp\left(-2c_0\hat{R} - \frac{\left(c + c_0 - \hat{N}\hat{R}\right)^2}{2\hat{N}}\right)$$

$$\times \int_0^{\hat{N}} \left(\frac{c_0^2\hat{N}}{2\pi\lambda^3\left(\hat{N} - \lambda\right)}\right)^{1/2} \exp\left(-\frac{\left(c\lambda - c_0\left(\hat{N} - \lambda\right)\right)^2}{2\hat{N}\lambda\left(\hat{N} - \lambda\right)}\right) d\lambda.$$

$$(9.97)$$

The λ integral can now be evaluated, by the change of variable to $w \equiv \left[c\lambda - c_0\left(\hat{N} - \lambda\right)\right] / \left(\hat{N}\lambda\left(\hat{N} - \lambda\right)\right)^{1/2}$. Then

$$\int_0^{\hat{N}} \left(\frac{c_0^2 \hat{N}}{2\pi \lambda^3 \left(\hat{N} - \lambda \right)} \right)^{1/2} \exp\left(-\frac{\left(c\lambda - c_0 \left(\hat{N} - \lambda \right) \right)^2}{2\hat{N}\lambda \left(\hat{N} - \lambda \right)} \right) d\lambda$$

(9.98)

$$= \frac{1}{\sqrt{2\pi}} \int_{-\infty}^{\infty} \left(1 - \frac{w}{(w^2 + 4cc_0/\hat{N})^{1/2}} \right) \exp\left(-\frac{w^2}{2} \right) dw = 1.$$

The resulting expression for $\hat{\rho}_N(C)$ is just Eq. (9.13) in the text.

Appendix 2: Chain Rule Convolution

Begin with the expression for the ratio,

$$J \equiv \frac{1}{\tau_E} \int_0^{\tau_E} d\tau \int_{-\infty}^{\infty} d\gamma \, \rho_1(\gamma_E - \gamma, \tau_E - \tau) \rho_E(\gamma, \tau) / \rho_1(\gamma_E, \tau_E).$$

(9.99)

After substituting for ρ_1 and ρ_E,

$$J = \frac{1}{\gamma_E} \int_{-\infty}^{\gamma_E} d\gamma \int_0^{\tau_E} d\tau \left(\frac{(\gamma_E - \gamma)^2 \tau_E}{2\pi \, (\tau_E - \tau)^3 \tau} \right)^{1/2} \left\{ \exp\left[-\left(\frac{\gamma}{\tau} - \frac{\gamma_E}{\tau_E} \right)^2 \middle/ 2 \left(\frac{1}{\tau} - \frac{1}{\tau_E} \right) \right] \right.$$

$$\left. - \exp\left[-\left(\frac{2\gamma_E - \gamma}{\tau} - \frac{\gamma_E}{\tau_E} \right)^2 \middle/ 2 \left(\frac{1}{\tau} - \frac{1}{\tau_E} \right) \right] \right\}.$$

(9.100)

Next change integration variable: for the first term inside the braces substitute $u = \gamma^2 \left(\tau^{-1} - \tau_E^{-1} \right)$, while for the second term substitute $u' = (2\gamma_E - \gamma)^2 \left(\tau^{-1} - \tau_E^{-1} \right)$. Then

$$J = \frac{1}{\gamma_E} \int_{-\infty}^{\gamma_E} d\gamma \left\{ \int_0^{\infty} du \frac{|\beta|}{\sqrt{2\pi \, u^3}} \exp\left(-\frac{(u + \beta)^2}{2u} \right) \right.$$

$$\left. - \int_0^{\infty} du' \frac{|\beta'|}{\sqrt{2\pi \, u'^3}} \exp\left(-\frac{(u' + \beta')^2}{2u'} \right) \right\},$$

(9.101)

where $\beta \equiv (\gamma - \gamma_E) \, \gamma \, / \tau_E$, $\beta' \equiv (\gamma_E - \gamma)(2\gamma_E - \gamma)/\tau_E$. But

$$\int_0^\infty du \frac{|\beta|}{\sqrt{2\pi \, u^3}} \exp\left(-\frac{(u+\beta)^2}{2u}\right) = \exp(-\beta - |\beta|), \qquad (9.102)$$

as listed, e.g., in *Mathematica* or derivable via the further change of variable $v = (u + \beta)/\sqrt{u}$. Finally, shift the γ integration to $\tilde{\gamma} = -\gamma$ for the first term in the braces and to $\tilde{\gamma}' = \gamma_E - \gamma$ for the second, revealing a major cancellation between the two terms while the un-cancelled part of the $\tilde{\gamma}$ integral becomes trivial. The result is just $J = 1$, QED.

Chapter 10
Play Strategies with Card Counting

10.1 Count-Dependent Playing Strategy

The analysis leading to Basic Strategy calculates the expected return, R, for a large number of possible playing strategies and identifies the one that maximizes it. The hand values s and s', at or above which Player should stand, are identified as well as the two-card values for which he should double down or split. Taken together, the values for standing, doubling and splitting—for each Dealer upcard—constitute a set of parameters that specify each possible strategy. The family of all possible strategies numbers in the millions, although only one is maximal. Designate the family as π.

The task here is to develop a methodology for finding the optimal π dependent on the true count(s). The task is complicated by the fact that π is a disparate collection of individual numbers, each of which is an integer and none larger than 20: the standard techniques of the calculus are not suited to discrete variables of this kind. Also, a computer simulation approach would be very cumbersome, since for each true count many distinct strategy sets would need to be computed and compared to find the optimum; and the related task of finding the optimal counting vector(s) would then require repeating the entire computation for each of many candidate vectors. Although neither approach would be successful by itself, in combination they can give at least a qualitative insight.

Begin the maximization analysis as if π were a set of continuous, rather than discrete, elements labeled with the index ν. For this warm-up exercise, make π an explicit variable of R; then the yield, generalizing Eq. (8.24), is

$$Y = \left\langle \left\langle B(\hat{\gamma}) \left\langle R_f(\mathbf{d}; \pi) \right\rangle_{\hat{\gamma}} \right\rangle \right\rangle. \tag{10.1}$$

Also designate π_0 as the optimal parameter set that maximizes Y at $\hat{\gamma} = 0$ (the analog of Optimal Basic Strategy). In addition, again recognize Δ^2 as small, as in Sect. 8.1.3, and truncate the Hermite expansion of Eq. (8.20) after $n = 2$. Then a

© Springer International Publishing AG, part of Springer Nature 2018

N. R. Werthamer, *Risk and Reward*, https://doi.org/10.1007/978-3-319-91385-8_10

series expansion in powers of $\delta\pi \equiv \pi - \pi_0$ gives

$$\langle R_f(\mathbf{d};\pi)\rangle_{\hat{\gamma}} \cong R^0 + \frac{1}{2}\delta\pi\,\delta\pi : \frac{\partial^2 R^0}{\partial\pi_0\partial\pi_0}$$

$$-\sum_\lambda \hat{\gamma}_\lambda \hat{\boldsymbol{\alpha}}_\lambda \cdot \hat{\nabla}\left(R^0 + \delta\pi \cdot \frac{\partial R^0}{\partial\pi_0}\right) \tag{10.2}$$

$$+\frac{1}{2}\sum_{\lambda,\lambda'}\left(\hat{\gamma}_\lambda\hat{\gamma}_{\lambda'} - \Delta^2\delta_{\lambda,\lambda'}\right)\hat{\boldsymbol{\alpha}}_\lambda\hat{\boldsymbol{\alpha}}_{\lambda'} : \hat{\nabla}\hat{\nabla}R^0$$

with $R^0 \equiv R_0(\mathbf{d}_0;\pi_0)$. Maximizing with respect to $\delta\pi$ leads to the optimal value

$$\langle R_f(\mathbf{d};\pi_0)\rangle_{\hat{\gamma}} = R^0 - \sum_\lambda \hat{\gamma}_\lambda\hat{\boldsymbol{\alpha}}_\lambda \cdot \hat{\nabla}R^0$$

$$+\frac{1}{2}\sum_{\lambda,\lambda'}\hat{\boldsymbol{\alpha}}_\lambda\hat{\boldsymbol{\alpha}}_{\lambda'} : \left[\left(\hat{\gamma}_\lambda\hat{\gamma}_{\lambda'} - \Delta^2\delta_{\lambda,\lambda'}\right)\hat{\nabla}\hat{\nabla}R^0 + \hat{\gamma}_\lambda\hat{\gamma}_{\lambda'}\mathbf{Q}\right],$$

$$\tag{10.3}$$

where $\mathbf{Q}$ involves derivatives of R with respect to π_0,

$$\mathbf{Q} \equiv \sum_{\nu,\nu'}\frac{\partial\hat{\nabla}R^0}{\partial\pi_{0,\nu}}\left(\frac{\partial^2 R^0}{\partial\pi_0\partial\pi_0}\right)^{-1}_{\nu,\nu'}\frac{\partial\hat{\nabla}R^0}{\partial\pi_{0,\nu'}}. \tag{10.4}$$

Furthermore, since $\hat{\mathbf{d}}\cdot\hat{\nabla} = 0$ as per Eq. (8.9), hence $\hat{\mathbf{d}}\cdot\mathbf{Q} = \mathbf{Q}\cdot\hat{\mathbf{d}} = 0$.

Next, some of this analysis is at least qualitatively correct in the actual situation that π is a set of discrete variables rather than continuous ones. The approach assumes that Eq. (10.3) is still appropriate, i.e., that $\langle R_f(\mathbf{d};\pi)\rangle_{\hat{\gamma}}$ is a quadratic form in $\sum_\lambda \hat{\gamma}_\lambda\hat{\boldsymbol{\alpha}}_\lambda$ with coefficients $\mathbf{Q}$ at second order, even though Eq. (10.4) is disregarded and instead $\mathbf{Q}$ is viewed as an array that must be determined computationally.

The assumption has already been validated numerically in Sect. 8.2. There, Fig. 8.1 displays results for the quantities

$$R_{\mathrm{opt}} \equiv R_0\left(\langle\mathbf{d}\rangle_{\hat{\gamma}};\pi\right) - R^0, \quad R_{\mathrm{bas}} \equiv R_0\left(\langle\mathbf{d}\rangle_{\hat{\gamma}};\pi_0\right) - R^0, \tag{10.5}$$

as a function of a single count $\hat{\gamma}$. It uses Eq. (8.11) for $\langle\mathbf{d}\rangle_{\hat{\gamma}}$ with a counting vector $\mathbf{v}_b$ as per Eq. (8.30); and for R_{opt}, π is continuously optimized with respect to $\hat{\gamma}$. The evidence of Fig. 8.1 is that both R_{opt} and R_{bas} can reasonably be approximated by cubic polynomials in true count, although Fig. 8.3 demonstrates the fine-grained kinkiness of R_{opt}.

Thus, returning to the computation of $\mathbf{Q}$, employ the following approach: while systematically varying $\hat{\boldsymbol{\alpha}}$ throughout its vector space, for each such $\hat{\boldsymbol{\alpha}}$ least-squares fit a cubic to $R_0\left(\langle\mathbf{d}\rangle_{\hat{\gamma}};\pi\right) - R_0\left(\langle\mathbf{d}\rangle_{\hat{\gamma}};\pi_0\right)$ as a function of $\hat{\gamma}$ and extract the coefficient of the quadratic term. This approach gives numerical values for all 55 independent elements of the $\mathbf{Q}$ array.

10.1.1 Counting Vector Optimal for Play Variation Alone

Having established the validity of Eq. (10.3) with computed coefficients, the next step is to maximize Y with respect to the counting vectors. In this subsection the task is simplified by assuming that no bet variation is carried out; the succeeding subsections explore varying bet and play strategies simultaneously.

If the bet is fixed, then the yield, combining Eqs. (10.1) and (10.3) together with Eq. (8.7) in the caret representation, reduces to

$$Y = B \left[R_0 + \frac{\Psi(F)}{52\,D} \frac{1}{2} \sum_{\lambda=1}^{\Lambda} \hat{\boldsymbol{\alpha}}_\lambda \hat{\boldsymbol{\alpha}}_\lambda : \mathbf{Q} \right], \tag{10.6}$$

where $\Psi(F)$, using Eq. (8.24), is

$$\Psi(F) \equiv 52D \left\langle\!\left\langle \Delta^2 \right\rangle\!\right\rangle = \frac{1}{F} \int_0^F df \frac{f}{1-f} = -1 + \frac{1}{F} \ln \frac{1}{1-F}, \tag{10.7}$$

and is plotted in Fig. 10.1.

The techniques of Chap. 8, especially its Appendix 2, in Chap. 8, are applicable to finding the stationary points of Y with respect to $\hat{\boldsymbol{\alpha}}_\lambda$. The solution is facilitated by

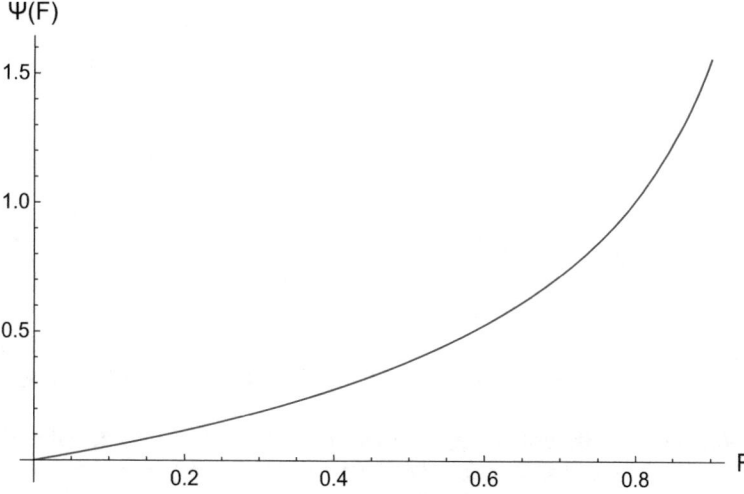

Fig. 10.1 The function $\Psi(F)$

introducing the eigenmodes (refer back to Appendix 2 in Chap. 8) of $\mathbf{Q}$,

$$\mathbf{Q} = \sum_{\kappa=1}^{9} Q_\kappa \mathbf{e}_\kappa \mathbf{e}_\kappa, \tag{10.8}$$

with the index $\kappa = 1$ labelling the largest of the non-vanishing eigenvalues Q_κ and $\kappa = 9$ the smallest; the 10th eigenvector, $\hat{d}$, has eigenvalue zero. It is easily shown that Y is stationary when each $\hat{\alpha}_\lambda$ aligns with an eigenvector $\mathbf{e}_\kappa$. Substitution into Eq. (10.6) shows that Y is maximized when the Λ largest of the eigenmodes are selected: the maximum Y is given by

$$Y = B\left[R_0 + \frac{\Psi(F)}{52\,D} \frac{1}{2} \sum_{\kappa=1}^{\Lambda} Q_\kappa \right]. \tag{10.9}$$

An important consequence of Eq. (10.9) is that the number of useful counting vectors is no longer confined to just one. Although the first vector picks up the largest contribution to Y, additional vectors add further contributions. However, if the largest eigenvalue Q_1 is much larger than the sum of all the others, then that eigenvector provides the dominant effect and the others are insignificant; the key measure is the ratio $Q_1/\mathrm{tr}\,\mathbf{Q}$, where the trace operation is defined as $\mathrm{tr}\,\mathbf{Q} \equiv \sum_{\kappa=1}^{9} Q_\kappa$. Furthermore, a relevant measure of the worth of an approximate counting vector is its "play correlation," $C_p \equiv \hat{\alpha}^* \cdot \mathbf{e}_1$, analogous to the betting correlation of Eq. (8.31).

Since the elements of $\mathbf{Q}$ are available numerically, it is straightforward to compute its eigenmodes and so assess the incremental yield from varying the play strategy. The result for the largest eigenvalue of $\mathbf{Q}$ is $Q_1 = 0.833$, while the trace is $\mathrm{tr}\,\mathbf{Q} = 1.256$. Thus the largest eigenmode accounts for just about 2/3 of the trace: that single counting vector realizes much (but certainly not nearly all) of the total yield improvement available from play variation. Additional numerical results, useful in other subsections, are

$$\mathbf{v}_b \cdot \mathbf{Q} \cdot \mathbf{v}_b = 0.556, \text{ and } |\mathbf{v}_b \cdot \mathbf{Q} \cdot (1 - \mathbf{v}_b \mathbf{v}_b)| = 0.241.$$

Next, the eigenvector $\mathbf{e}_1$ and the optimal counting vector for play variation $\alpha^*(j) \equiv e_1(j)/\hat{d}(j)$, are listed in Table 10.1; the second row is also shown in Table 5.1. As noted in Sect. 5.1.1, where it is labeled vp, this counting vector is similar to the optimal bet variation counting vector of Table 8.2, except for the striking reversal of sign for aces ($j = 1$) and the interchange of $j = 6$ with $j = 8$. Thus, rather than the two counting vectors being orthogonal, they have a strong correlation: $\mathbf{v}_b \cdot \mathbf{e}_1 = 0.833$.

Although α^* has irrational elements and so is not practical for actual counting, several approximations with integer or half-integer elements suggest themselves, analogous to those of Tables 3.1 and 3.2; both α^* and its approximation are

Table 10.1 Optimal counting vector for play variation

j	1	2	3	4	5	6	7	8	9	10
$e_1(j)$	+0.10	+0.09	+0.16	+0.30	+0.42	+0.06	+0.25	+0.18	−0.03	−0.77
$\alpha^*(j)$	+0.37	+0.33	+0.57	+1.10	+1.52	+0.22	+0.88	+0.66	−0.12	−1.38
$\alpha^*_{HO}(j)$	0	0	0	+1	+1	0	+1	+1	0	−1

balanced. The simplest, analogous to the Hi-Opt scheme but with the switch of $j = 3, 6$ with $j = 7, 8$ is also shown in Table 10.1, designated α^*_{HO}, and as the "simplified" row of Table 5.1. It has a play correlation of $C_p = 0.95$. The Hi-Opt vector itself has $C_p = 0.88$, larger than the $C_p = 0.80$ of the Hi-Lo vector and even the 0.83 of v_b; the higher play correlation of the Hi-Opt vector confirms the claim of its proponents that it does a better job than other simple approximations in taking advantage of play variation.

When counting with the vector α^* and adjusting play strategy, the yield improves according to Eq. (10.9); numerically,

$$Y/B = R_\infty + (0.0067 + 0.0080\Psi(F))/D. \tag{10.10}$$

For casinos that reshuffle frequently, e.g., after only half the pack is dealt, $\Psi \approx 0.4$; at the other end of the range, few casinos will put off reshuffling until the penetration is as much as $F = 0.8$, where $\Psi \approx 1$. More typical might be the intermediate value, $\Psi \approx 0.7$. The result is that play variation improves the yield ratio only slightly, +0.002 at most, for $D = 4$ or more; it does not in itself shift the yield even close to positive. For $D = 1$, however, where the Optimal Basic Strategy return is already positive, play variation boosts the yield ratio by perhaps another +0.006.

10.1.2 Single Counting Vector Optimal for Bet and Play Together

It is disappointing, but not surprising, that varying the play strategy, based on the true count with a well-chosen counting vector, provides only a slight improvement in yield. This outcome is similar to the negligible improvement found in Sect. 2.1 from variations in optimal play with the number of decks or the composition of the hand. It seems that the yield is quite insensitive to changes in optimal play strategy arising from small changes in expected return. Nevertheless, pursuing the possibility to its logical end, consider how bet and play variations interact with each other in determining the best choice of counting vector(s) and the resulting yield.

The detailed analysis differs depending on whether only a single vector is considered, optimized to take best advantage of bet and play strategies together, or two vectors, one specific for betting and the other for playing. This subsection focuses on the former case, the next on the latter. Combine expressions 8.32 and

10.5 (generalized to a varying bet size) to obtain

$$Y = Y_0 + Y_1 \hat{\boldsymbol{\alpha}} \cdot \mathbf{v}_b + \hat{\boldsymbol{\alpha}}\hat{\boldsymbol{\alpha}} : \left(Y_{21} \hat{\nabla}\hat{\nabla} R^0 + Y_{22} \mathbf{Q} \right) \Big/ 2, \tag{10.11}$$

where $Y_1 \equiv \langle\langle B\left(\hat{\gamma}\right)\hat{\gamma}\rangle\rangle \big| \hat{\nabla} R^0 \big|$, $Y_{21} \equiv \langle\langle B\left(\hat{\gamma}\right)\left(\hat{\gamma}^2 - \Delta^2\right)\rangle\rangle$, $Y_{22} \equiv \langle\langle B\left(\hat{\gamma}\right)\hat{\gamma}^2\rangle\rangle$. Leaving aside the Y_{21} term, the maximization of Y with respect to $\hat{\boldsymbol{\alpha}}$ is again facilitated by using the eigenmodes of $\mathbf{Q}$, leading to

$$\alpha_\kappa \equiv \hat{\boldsymbol{\alpha}} \cdot \mathbf{e}_\kappa = Y_1 v_\kappa / (L - Y_{22} Q_\kappa). \tag{10.12}$$

The Lagrange parameter L is determined by the normalization of $\hat{\boldsymbol{\alpha}}$, and so satisfies the implicit equation

$$1 = \sum_\kappa \left[Y_1 v_\kappa / (L - Y_{22} Q_\kappa) \right]^2. \tag{10.13}$$

Equation (10.13) has at least two solutions, as can be seen by graphing its right-hand side (rhs) as a function of L/Y_{22} and examining its crossings with 1, the left-hand side (schematically, Fig. 10.2).

Since the rhs tends to zero for L tending to $\pm\infty$, and tends to infinity (a pole) for L/Y_{22} tending to each Q_κ, there are always solutions for $L/Y_{22} > Q_9$ and for $L/Y_{22} < Q_0 = 0$. The first of these maximizes Y, the second minimizes it. There may or may not also be solutions for L intermediate between two consecutive poles, depending on the depth of the minimum between those poles; the depth, in turn, is

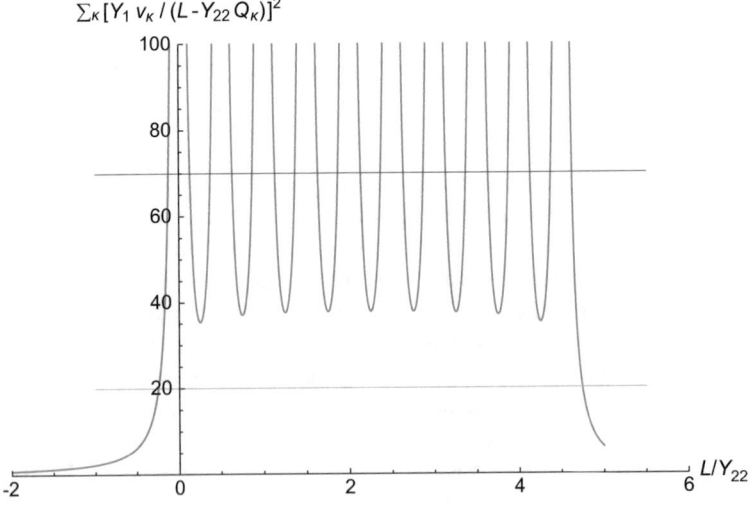

Fig. 10.2 Schematic graphical solutions to the eigenvalue equation

controlled by the ratio $\eta \equiv Y_{22}/Y_1$. If the ratio is large (for example, the lhs given by the red line), then solutions exist for L near every pole; in the limit, these tend to the solutions of the previous section. If the ratio is small (for example, the lhs given by the green line), then no intermediate solutions exist; in the limit, the two "outer" solutions tend to those of Chap. 8.

In principle, $\hat{\boldsymbol{\alpha}}$ could be computed, for any given η, using the numerical eigenmodes of $\mathbf{Q}$ and Eqs. (10.11) and (10.12). But in practice η is small when bet variation is carried out, so insight is gained by proceeding further analytically. A series expansion of these relations to first order in powers of η finds

$$\hat{\boldsymbol{\alpha}} \equiv \mathbf{v}_b(1 - |\boldsymbol{\alpha}_\perp|^2)^{1/2} + \boldsymbol{\alpha}_\perp,$$

$$\boldsymbol{\alpha}_\perp \cong \eta \mathbf{v}_b \cdot \mathbf{Q} \cdot (1 - \mathbf{v}_b \mathbf{v}_b).$$

(10.14)

The presence of play variation shifts the optimal counting vector away from $\mathbf{v}_b$, by adding a small component perpendicular to it while shrinking the magnitude along it to preserve the normalization.

With the counting vector optimized, the resulting yield can be obtained by substituting back into Eq. (10.11); through second order in η and using the numerical results of the previous section,

$$Y \cong Y_0 + Y_1 + \left(Y_{22}\mathbf{v}_b \cdot \mathbf{Q} \cdot \mathbf{v}_b + Y_1|\boldsymbol{\alpha}_\perp|^2\right)\Big/ 2$$

$$= Y_0 + Y_1\left(1 + .28\eta + .03\eta^2\right).$$

(10.15)

The first three terms of this last expression are together just the yield if $\hat{\boldsymbol{\alpha}}$ were not shifted from $\mathbf{v}_b$ as per Eq. (10.14): the shift increases Y only to second order in η—the last term of Eq. (10.15)—an increase that is essentially negligible. Thus, even if both play and bet size are varied, to a good approximation a single counting vector remains fixed at its optimum with respect to bet variation alone; and this conclusion still holds even with the more elaborate analysis that includes the Y_{21} term. Furthermore, the added yield from play variation even without the shift in counting vector (the third term in Eq. (10.15)) is almost exactly 2/3 of that available if the play variation were at its maximum, $C_p = 1$; maneuvers to increase yield still more have little left to gain.

10.1.3 Two Distinct Counting Vectors

As mentioned earlier, the strategy of varying both bet and play suggests the possibility of using two different counting vectors, one for choosing the bet size and the other for choosing the play. It seems intuitive that the yield, Y, would be optimized by selecting $\mathbf{v}_b$ as the counting vector for bet size and $\mathbf{e}_1$ for play; and that the resulting yield would be $Y = Y_0 + Y_1 + Y_{22} \, Q_1/2$, an improvement over that of Eq. (10.15) unless η is quite large (> 4.8). Although these simple suppositions are indeed correct, deriving them requires some care.

The mathematical challenge is that $\mathbf{v}_b$ and $\mathbf{e}_1$ are not orthogonal: in fact, as seen earlier, they are substantially correlated. Thus they will not each align with a distinct optimal orthonormal vector $\hat{\boldsymbol{\alpha}}_\lambda$. Structure this problem by designating $\hat{\boldsymbol{\alpha}}_2$ as the counting vector for play variation, and allowing the counting vector for bet variation to be some linear combination of the $\hat{\boldsymbol{\alpha}}_\lambda$. If so, then $\langle\!\langle B\hat{\gamma}_\lambda \rangle\!\rangle = Y_1 \theta_\lambda$ where, except for the normalization $\sum_\lambda \theta_\lambda^2 = 1$, the θ_λ remain to be determined. The yield in this situation is

$$Y = Y_0 + Y_1 \sum_\lambda \theta_\lambda \hat{\boldsymbol{\alpha}}_\lambda \cdot \mathbf{v}_b + Y_{22}\hat{\boldsymbol{\alpha}}_2\hat{\boldsymbol{\alpha}}_2 : \mathbf{Q}/2. \tag{10.16}$$

Thus Y is stationary with respect to the $\hat{\boldsymbol{\alpha}}_\lambda$ provided

$$Y_1\theta_\lambda\hat{\boldsymbol{\alpha}}_\lambda \cdot \mathbf{v}_b + Y_{22}\delta_{\lambda,2}\mathbf{Q} \cdot \hat{\boldsymbol{\alpha}}_2 = \sum_{\lambda,\lambda'} L_{\lambda,\lambda'}\hat{\boldsymbol{\alpha}}_{\lambda'}. \tag{10.17}$$

Setting $\hat{\boldsymbol{\alpha}}_2 = \mathbf{e}_1$, Eq. (10.17) separates into the two conditions

$$\begin{aligned} Y_1\theta_1\mathbf{v}_b &= L_{11}\hat{\boldsymbol{\alpha}}_1 + L_{12}\mathbf{e}_1, \\ Y_2\theta_2\mathbf{v}_b + Y_2Q_1\mathbf{e}_1 &= L_{12}\hat{\boldsymbol{\alpha}}_1 + L_{22}\mathbf{e}_1. \end{aligned} \tag{10.18}$$

Equations (10.18) are internally consistent only if $L_{11}\theta_2 = L_{12}\theta_1$ and $L_{11}(L_{22} - Y_2Q_1) = (L_{12})^2$. The remaining elements of $L_{\lambda,\lambda'}$ are fixed from the constraints that $\hat{\boldsymbol{\alpha}}_1$ is normalized and orthogonal to $\mathbf{e}_1$; thus $L_{12} = Y_1\theta_1 \cos\varphi$ where $\cos\varphi \equiv \mathbf{v}_b \cdot \mathbf{e}_1$, and $L_{11} = Y_1\theta_1 \sin\varphi$. Together, these determine the θ_λ: namely $\theta_1 = \sin\varphi$ and $\theta_2 = \cos\varphi$. These results, in turn, lead to

$$\hat{\boldsymbol{\alpha}}_1 = \mathbf{v}_b \csc\varphi - \mathbf{e}_1 \cot\varphi. \tag{10.19}$$

Finally, using Eqs. (10.18), the counting vector for bet variation $\sum_\lambda \theta_\lambda\hat{\boldsymbol{\alpha}}_\lambda$ becomes just $\mathbf{v}_b$, as intuition suggested. Furthermore, the stationary value of Y, by substituting into Eq. (10.15), also matches expectation; substituting the numerical result for Q_1 leads to $Y = Y_0 + Y_1(1 + 0.42\eta)$.

10.1.4   Insurance with Variable Betting

Section 7.1 discussed insurance for the Basic Strategy Player, who does not count or vary his bets, and showed that the expected return from insurance is negative. However, the return does become positive for large enough positive true counts.

Analysis is needed to determine whether Player can gain from accepting insurance and under what circumstances. Based on Eqs. (7.39) and (8.12), the expected return from insurance, conditional on a true count γ, is

$$\langle R_I \rangle_\gamma = \langle\langle d(1)(3\,d(10) - 1)/2\rangle\rangle_\gamma.$$ (10.20)

The simplification adopted by most authorities is to replace this with

$$\langle R_I \rangle_\gamma^0 \equiv \frac{1}{2}\langle\langle d(1)\rangle\rangle_\gamma \left(3\langle\langle d(10)\rangle\rangle_\gamma - 1\right)$$

$$= \frac{1}{338}\left(1 - \frac{\alpha(1)\gamma}{52}\right)\left(-1 - \frac{12\alpha(10)\gamma}{52}\right);$$ (10.21)

this return is positive for $\gamma \geq \gamma_I^0 \equiv -52/12\alpha(10), \approx 4.053$ with the counting vector $\alpha(10) = -1.0692$ optimal for betting. Accepting insurance for true counts in this range, the added yield becomes

$$Y_I = \langle\langle B(\gamma)\langle R_I\rangle_\gamma \ominus \left(\langle R_I\rangle_\gamma\right)\rangle\rangle$$

$$\cong \langle\langle B(\gamma)\langle R_I\rangle_\gamma^0 \ominus \left(\gamma - \gamma_I^0\right)\rangle\rangle.$$ (10.22)

Although the value of the simplified expression is small (depending specifically on the several parameters relevant to counting: the penetration, number of decks, and bet ramp), it is clearly positive and comparable to the yield gain from other count-dependent play decisions. Authors who discuss count-dependent play recommend including this insurance recipe.

However, Eq. (10.20) can in fact be evaluated exactly, using Eq. (8.20), as

$$\langle R_I \rangle_\gamma = \frac{1}{338}\left[\left(1 - \frac{\alpha(1)\gamma}{52}\right)\left(-1 - \frac{12\alpha(10)\gamma}{52}\right) - 12\Delta^2\alpha(1)\alpha(10)\right].$$ (10.23)

This more general expression can be computed straightforwardly and, with the optimal counting vector for betting, results in yield gain that is smaller than with Eq. (10.21) by roughly 20% to 25%, depending on the parameters mentioned above. The key cause is that the threshold for positive $\langle R_I\rangle_\gamma$, based on Eq. (10.23) and a counting vector for betting, increases monotonically from γ_I^0 with increasing depth. Thus, at depths approaching the penetration, the threshold becomes so high that it has little probability of being reached: the opportunities for insurance actually shrink as the pack is dealt, rather than expand. Conversely, retaining the γ_I^0 threshold at all depths imposes negative expectation bets at larger depths and so performs still worse, by another 10% or so. A better choice of an approximate, depth-independent true count threshold is around 4.5 for four decks, 5.5 for one deck, which perform nearly as well as the exact depth-dependent threshold.

Griffin (1999), p. 71 and Wong (1994), p. 54, among others, note that a second counting vector, optimized and used exclusively for the insurance decision, might improve the yield vs. using a counting vector designed for betting. Working within the simplified Eq. (10.21), they recognize that the (normalized) second vector which produces the lowest true count threshold and most steeply increasing $\langle R_I \rangle_\gamma^0$, and hence the greatest yield, is $\alpha(10) = -3/2$, $\alpha(j \neq 10) = 2/3$; this is much like Thorp's Tens-count. This lowest threshold then is $\gamma_I^0 = 52/18 = 2.889$, well below that for the Hi-Lo vector.

Considering the exact insurance return, Eq. (10.23), and maximizing its yield, shows that the threshold for positive $\langle R_I \rangle_\gamma$ shifts downward with depth (rather than upward, as it does with the bet counting vector, because of the change in sign of $\alpha(1)$) from γ_I^0, its value at zero. For Players who do not wish to accurately track the threshold down with depth, the best choice of a depth-independent threshold, with only slightly diminished performance, would correspond to an intermediate depth: a true count of about 2.7 for four decks, 2.1 for one deck. But even with such a second count, the insurance yield is still only improved by roughly 50% over that with just a bet vector, sufficiently modest that the additional counting effort seems unwarranted.

Strictly speaking, though, this Tens vector is the optimal counting vector only at or near zero depth, where the insurance option has very low likelihood. The optimal vector shifts with increasing depth, becoming noticeably different from the Tens (although its elements are no longer simple rationals, and so are impractical to use). Nevertheless, this refinement similarly makes only a slight improvement in yield. The bottom-line recommendation, reiterating Sect. 5.1, is to decline insurance unless the true count (with a bet counting vector such as Hi-Lo) exceeds roughly +5.

10.2   Counter Basic Strategy

This topic is the last in our book, but not at all the least significant. Recall that Basic Strategy is the simple way to play a hand that maximizes the expected return at the reshuffle round. Basic Strategy also maximizes the yield for a Player who does not count or vary his bets. For a card counter, though, the previous section considered the adjustment of play parameters depending on the true count, resulting in a complicated set of strategy indices and an only slightly improved yield. But if Player counts and varies his bets accordingly, a different approach to play variation suggests itself: a count- *independent* play strategy that maximizes the   *yield*, the bet-weighted average of expected return over all the rounds in a shoe rather than just at zero depth. Such a strategy, termed Counter Basic Strategy (CBS) by Marcus (2007), is no more complex than Basic Strategy itself, yet still captures much of the (modest) yield improvement possible from the strategy indices. The analysis given here follows Werthamer (2007) although, since it has many steps and is computationally intensive, only a sketch is given here. Its performance and the simple recipe for its use are detailed in Sect. 5.2.

Section 8.2 developed the Hermite expansion to extend the expected return to nonzero depths. In that analysis, the play parameters were implicit, fixed, and given by Basic Strategy, which maximizes the expected return at zero depth and zero true count. Here, though, consider those strategy parameters π_c that maximize the expected return at zero depth and at an as-yet-unspecified true count $\gamma_c > 0$. Then if the π_c are taken as independent of depth and used throughout the shoe, the yield will be

$$Y(\gamma_c) = \left\langle\!\left\langle B(\gamma) \left\langle R_f(\mathbf{d}; \pi_c) \right\rangle_\gamma \right\rangle\!\right\rangle, \tag{10.24}$$

as per Eq. (8.24). But since the bet sizes $B(\gamma)$ vary with true count and have a distribution that is peaked at some positive value of γ, the yield should be a maximum when γ_c lies somewhere near that peak. Label that true count γ^*, label the associated play parameters π^*, and label the resulting yield

$$Y^* = Y(\gamma^*) = \left\langle\!\left\langle B(\gamma) \left\langle R_f(\mathbf{d}; \pi^*) \right\rangle_\gamma \right\rangle\!\right\rangle. \tag{10.25}$$

The true count γ^* is defined as that which maximizes $Y(\gamma^*)$.

Furthermore, this expression becomes simpler via the Hermite expansion of Eq. (8.20):

$$Y^* = \sum_{n=0}^{\infty} (1/n!) \left\langle\!\left\langle B(\gamma) \Delta^n H_n(\hat{\gamma}/\Delta) \right\rangle\!\right\rangle \left(-\hat{\alpha} \cdot \hat{\nabla}\right)^n R_0(\mathbf{d}_0; \pi^*)$$

$$= \sum_{n=0}^{\infty} c_n^* \left\langle\!\left\langle B(\gamma) \Delta^n H_n(\hat{\gamma}/\Delta) \right\rangle\!\right\rangle. \tag{10.26}$$

In this form the coefficients c_n^*, which contain the play parameters, can be generated just from the return at the reshuffle round, and independent of the bet function used. The return's depth dependence, required for the yield, is then contained solely within the Hermite polynomials.

The computations to implement this analytical program are lengthy, comprising multiple steps. In outline:

1. Construct a code for the expected return from a pack with D decks at zero depth and zero true count. The play parameters here are those that maximize that reshuffle return, the same as $\pi(0)$ in Sect. 7.2.
2. Generalize the code to a non-vanishing true count γ, by using the card probabilities appropriate to that count: notationally, $\langle \mathbf{d} \rangle_\gamma$ rather than $\mathbf{d}_0$. Compute $R_0(\langle \mathbf{d} \rangle_\gamma; \pi)$ for various π, so that maximizing it with respect to π at each true count γ gives the parameters $\pi(\gamma)$ and the resulting expected return $R_0(\langle \mathbf{d} \rangle_\gamma; \pi(\gamma))$. It is just this computation that generates the curves of Figs. 8.1 and 8.2.

3. Since $R_0\left(\langle\mathbf{d}\rangle_\gamma; \boldsymbol{\pi}(\gamma)\right)$ deviates from linearity in γ as displayed in those figures, least–squares fit its computed results with a low-order polynomial in γ; as in Chap. 8, quite good accuracy (to about 3 significant figures throughout) is obtained with a cubic without DAS, a quartic with DAS. Then transcribe the four/five c_n coefficients from the fit into the corresponding Hermite polynomial expansion Eq. (8.20), which extends the return to finite depth. The truncated Hermite series gives an excellent numerical approximation to $\left\langle R_f(\mathbf{d}; \boldsymbol{\pi})\right\rangle_\gamma$. So far this is just the program of Chap. 8.

4. Now select several positive trial values γ_c for the true count, which so far has been arbitrary. For each, generate $\boldsymbol{\pi}(\gamma_c)$ from the results of step (2), i.e., by maximizing the expected return at that count. Then, for this fixed set of play parameters, compute $R_0\left(\langle\mathbf{d}\rangle_\gamma; \boldsymbol{\pi}(\gamma_c)\right)$ as a function of γ. A truncated Hermite polynomial expression is generated just as in the previous paragraph, approximating the corresponding $\left\langle R_f\left(\mathbf{d}; \boldsymbol{\pi}(\gamma_c)\right)\right\rangle_\gamma$; this quantity is maximal, of course, only at $\gamma = \gamma_c$.

5. Adopt one of the optimal bet ramps $B(\gamma)$ developed in Sect. 9.1; multiply it with the expected return approximation from step (4); and then use Eq. (8.24) to compute the yield Y_c corresponding to each γ_c, by integrating over γ weighted by $p(\gamma)$ and by averaging over depth.

6. The set of trial yield values Y_c are interpolated and the maximum is extracted; at that maximum, the resulting $\boldsymbol{\pi}^*$ represents the desired Counter Basic Strategy, and Y^* the corresponding yield. Note, however, that CBS is dependent on the particular choice of bet strategy, as well as on the counting vector and the penetration. A representative case is detailed in Table 5.4.

A project for the future is to assemble these steps into a single, unified computational code, even if complex and lengthy to execute, which could produce results for a variety of playing conditions and other input parameters.

References and Bibliography

Anonymous, Ashley, Elizabeth, and Maria, "The Secret Diary of an All-Girl Blackjack Team", CreateSpace Publishing, (2015)

Baldwin, Roger, Wilbert E. Cantey, Herbert Maisel, and James P. McDermott, "The Optimum Strategy in Blackjack", Journal of the American Statistical Association **51**, 429–439 (1956)

Black, Fischer and Myron Scholes, "The Pricing of Options and Corporate Liabilities", Journal of Political Economy **81**, 637–654 (1973)

Braun, Julian H., "How to Play Winning Blackjack", Data House Publishing, Chicago, (1980)

Carlson, Bryce, "Blackjack for Blood", Ingram Publishing, 3rd edition, (2017)

Cox, J.C., S.A. Ross, and M. Rubenstein, "Option Pricing: a Simplified Approach", Journal of Financial Economics **7**, 229–263 (1979)

Epstein, Richard A., "The Theory of Gambling and Statistical Logic", Academic Press, New York, revised edition (1995)

Epstein, Richard A., "The Theory of Gambling and Statistical Logic", Academic Press, New York, 2nd edition (2009)

Ethier, S.N., "The Kelly System Maximizes Median Fortune", Journal of Applied Probability **41**, 230–6 (2004)

Griffin, Peter A., "The Theory of Blackjack", Huntington Press, Las Vegas, 6th edition (1999)

Grosjean, James, "Beyond Counting", RGE Publishing, Oakland (2000)

Harris, B., "The Theory of Optimal Betting Spreads"; Janacek, K., "Theory of Optimal Betting"; Yamashita, W., "Optimal Betting Strategy": all at www.bjmath.com/bjmath/Betsize (1997)

Hull, J.C., "Options, Futures and Other Derivatives", Prentice Hall, Englewood Cliffs, NJ (2002)

Ingersoll, J. E., Jr., "Theory of Financial Decision Making", Rowman & Littlefield, New York (1987)

Ito, C., "On Stochastic Differential Equations", Memoirs of the American Mathematical Society **4**, 1–51 (1951)

Kelly, John L., Jr., "A New Interpretation of Information Rate", System Technical Journal **35**, 917–926 (1956)

MacLean, Leonard C., Edward O. Thorp and William T. Ziemba, "The Kelly Capital Growth Investment Criterion: Theory and Practice", World Scientific (2010)

Manson, A.R., A.J. Barr, and J.H. Goodnight, "Optimum Zero-Memory Strategy and Exact Probabilities for 4-Deck Blackjack", The American Statistician **29**, 84–88 (1975)

Marcus, H.I., "New Blackjack Strategy for Players who Modify their Bets Based on the Count", in Ethier, S.N. and W.R. Eadington (eds.), "Optimal Play: Mathematical Studies of Games and Gambling", Institute for the Study of Gambling and Commercial Gaming, University of Nevada, Reno (2007)

Marsden, Jerrold E. and Anthony J. Tromba, "Vector Calculus", Freeman, 5th edition (2007)

© Springer International Publishing AG, part of Springer Nature 2018

N. R. Werthamer, *Risk and Reward*, https://doi.org/10.1007/978-3-319-91385-8

Maslov, Sergei and Y.-C. Zhang, "Optimal Investment Strategy for Risky Assets", International Journal of Theoretical and Applied Finance **1**, 377–387 (1998); Matteo Marsili, Sergei Maslov and Y.-C. Zhang, Physica A **253**, 403–418 (1998)

Merton, Robert C., "Theory of Rational Option Pricing", Bell Journal of Economics & Management Science **4**, 141 (1973)

Mezrich, Ben, "Bringing Down the House", Free Press, New York (2002)

Mezrich, Ben, "Busting Vegas", William Morrow, New York (2006)

Morse, P.M., H. Feshbach, "Methods of Theoretical Physics", McGraw-Hill, New York (1953)

Nersesian, Robert A., "The Law for Gamblers: A Legal Guide to the Casino Environment", Huntington Press (2016)

Poundstone, William, "Fortune's Formula", Hill and Wang, New York (2005)

Revere, Lawrence, "Playing Blackjack as a Business", Carol Publishing, Secaucus NJ, revised edition (1980)

Samuelson, Paul A., "Proof that Properly Anticipated Prices Fluctuate Randomly", Industrial Management Review **6**, 41 (1965)

Scarne, John, "Scarne's Complete Guide to Gambling", Simon & Schuster, New York (1961)

Schlesinger, Don, "Blackjack Attack: Playing the Pros' Way", RGE Publishing, North Las Vegas, 3rd edition (2005)

Scott, Jean, "The Frugal Gambler", Huntington Press, Las Vegas, 2nd edition (2005)

Shores, Thomas S., "Applied Linear Algebra and Matrix Analysis", Springer, Berlin (2007)

Sileo, Patrick, "The Evaluation of Blackjack Games Using a Combined Expectation and Risk Measure", in Eadington, W.R. and J.A. Cornelius (eds.), "Gambling and Commercial Gaming", University of Nevada, Reno (1992)

Snyder, Arnold, "Blackbelt in Blackjack", Cardoza Publishing, Las Vegas, 3rd edition (2005)

Snyder, Arnold, "The Big Book of Blackjack", Cardoza Publishing, Las Vegas (2006)

Thorp, Edward O., "Beat the Dealer", Blaisdell, New York, revised edition (1966)

Thorp, Edward O., "Does Basic Strategy Have the Same Expectation for Each Round?", in Vancura, O., J.A. Cornelius, and W.R. Eadington (eds.), "Finding the Edge: Mathematical Analysis of Casino Games", University of Nevada, Reno (2000)

Tilton, Nathaniel, "The Blackjack Life", Huntington Press, Las Vegas (2012); for other memoirs see Scoblete, Frank, " I Am a Card Counter: Inside the World of Advantage Play Blackjack ", (2014); Anonymous, " The Secret Diary of an All - Girl Blackjack Team ", (2015)

Vancura, Olaf, and Ken Fuchs, "Knock-Out Blackjack", Huntington Press, Las Vegas, 3rd edition (2016)

Werthamer, N. Richard, "Optimal Betting in Casino Blackjack", International Gambling Studies **5**, 253 (2005)

Werthamer, N. Richard, "Optimal Betting in Casino Blackjack II: Back-counting", International Gambling Studies **6**, 111 (2006)

Werthamer, N. Richard, "Basic Strategy for Card Counters: An Analytic Approach", in Ethier, S.N. and W.R. Eadington (eds.), "Optimal Play: Mathematical Studies of Games and Gambling", Institute for the Study of Gambling and Commercial Gaming, University of Nevada, Reno (2007)

Werthamer, N. Richard, "Optimal Betting in Casino Blackjack III: Table-hopping", International Gambling Studies **8**, 63 (2008)

Wong, Stanford, "Professional Blackjack", Pi Yee Press, Las Vegas, 5th edition (1994); "Blackjack Secrets", Pi Yee Press, Las Vegas (1994)

Zender, Bill, "Casino-ology: The Art of Managing Casino Games", Huntington Press, Las Vegas (2008)

Index

© Springer International Publishing AG, part of Springer Nature 2018
N. R. Werthamer, *Risk and Reward*, https://doi.org/10.1007/978-3-319-91385-8

Printed in the United States
By Bookmasters